The Age of Sharing

For Etty

The Age of Sharing

Nicholas A. John

polity

First published in 2017 by Polity Press

Polity Press
65 Bridge Street
Cambridge CB2 1UR, UK

Polity Press
350 Main Street
Malden, MA 02148, USA

ISBN-13: 978-0-7456-6250-3
ISBN-13: 978-0-7456-6251-0(pb)

A catalogue record for this book is available from the British Library.

Library of Congress Cataloging-in-Publication Data

Names: John, Nicholas A., author.
Title: The age of sharing / Nicholas A. John.
Description: Malden, MA : Polity, 2016. | Includes bibliographical references and index.
Identifiers: LCCN 2016015484 (print) | LCCN 2016033678 (ebook) | ISBN 9780745662503 (hardback) | ISBN 9780745662510 (paperback) | ISBN 9781509512287 (Mobi) | ISBN 9781509512294 (Epub)
Subjects: LCSH: Sharing. | Sharing–Social aspects. | Mass media–Social aspects. | BISAC: LANGUAGE ARTS & DISCIPLINES / Communication Studies.
Classification: LCC BF575.S48 .J64 2016 (print) | LCC BF575.S48 (ebook) | DDC 302/.1–dc23
LC record available at https://lccn.loc.gov/2016015484

Typeset in 10.5 on 12 pt Sabon
by Toppan Best-set Premedia Limited
Printed and bound in Great Britain by CPI Group (UK) Ltd, Croydon

For further information on Polity, visit our website: politybooks.com

Contents

Figures

Preface

Parts of this book have been published before, but none of those publications about sharing are reproduced here *in toto*. Parts of my article, 'Sharing and Web 2.0: The Emergence of a Keyword' (John, 2013a), appear in Chapter 3, and parts of 'File sharing and the History of Computing: Or, Why File Sharing is Called "File Sharing"' (John, 2014) appear in Chapter 6. Some of my first efforts at a theoretical analysis of sharing today (John, 2013b) appear throughout. The original research carried out for Chapters 2 and 4 and the second part of Chapter 6 has not been previously published.

Before their inclusion here, various ideas in this book had been presented at conferences, especially those of the Association of Internet Researchers and the International Communication Association.

This research was supported by The Israel Science Foundation (grant No. 38/14).

Acknowledgements

ı

I arrived at the Department of Communication and Journalism at the Hebrew University of Jerusalem in 2010 to carry out postdoctoral research into globalization and multichannel television, but was soon diverted by the observation that the word 'sharing' seemed to be all over the internet, and that this seemed vaguely interesting. I laid out some of my preliminary thoughts to Paul Frosh and Limor Shifman and they immediately started encouraging me to develop them and helped me to do so. Paul helped me with the process of finding a publisher, and Limor's input to my first publication about sharing was invaluable. You would not be reading these words if not for their support, advice and intellectual contributions. A number of other people were remarkably supportive and helpful as this project got off the ground, nudging me in new and interesting directions, and listening to my efforts to formulate and formalize my thoughts: Menahem Blondheim, Elihu Katz, Tamar Leibes, Amit Pinchevski and Keren Tenenboim-Weinblatt. For a year I shared an office with Benjamin Peters, which was both fun (obviously) and intellectually rewarding. Ben and I had many terrific conversations about sharing, and I'm grateful to him for them. While doing the research for this book, I was lucky enough to be appointed to the faculty of the Department of Communication and Journalism at the Hebrew University of Jerusalem. I thank my colleagues for the fantastic intellectual environment that they provide. Thanks also to Tzlil Sharon, my research assistant, for her dedicated assistance. For his unceasing encouragement and support, I particularly want to acknowledge the role played by Zohar Kampf as my sharing project took off and this book started to take form. His belief in my work was contagious, and I doubt whether the project would have got this far without it. I am extremely grateful to him.

Outside of my department, many other people have provided feedback, comments and suggestions. I would like to single out Russell Belk, whose intellectual generosity is inspirational. Russ's influence is felt throughout the book, perhaps most strongly at those points where my views diverge from his. I would also like to acknowledge my erstwhile supervisor, Eva Illouz, who helped me to sharpen the contribution of the book as a whole before I started writing it, and provided the impetus to the questions discussed in Chapter 6. The time I spent at the Department of Media and Communications at the LSE was invaluable, and I thank Robin Mansell, Bart Cammaerts, Shani Orgad and Alison Powell for their thoughts and wisdom. Finally, I wish to thank Fred Turner, Jenny Kennedy, Michal Hamo and also the attendees at Benjamin Peters' digital keywords retreat for their comments, input and discussions about the issues and work presented in this book.

I am indebted to the institutions that hosted me while I got this project under way as a postdoctoral researcher at the Department of Communication at the Hebrew University of Jerusalem, and then at the Department of Communication at Ben-Gurion University of the Negev, and as a Visiting Scholar at the Department of Media and Communications at the LSE. I am further grateful for the funding I received to support my research, in particular the fellowship from the Lady Davis Fellowship Trust at the Hebrew University and the scholarship from the Kreitman School of Advanced Graduate Studies at Ben-Gurion University. I also deeply thank the Israel Science Foundation for its support.

The people at Polity have been a pleasure to work with. The peerless Andrea Drugan steered me through the proposal stage and set me on my way, while Lauren Mulholland and Elen Griffiths were there to guide me as the text was being written. I thank Elen, Ellen MacDonald-Kramer, Helen Gray and all at Polity for bringing this book into being.

Aram Sinnreich tutted and shook his head when I told him the title I originally had in mind for this book, gave the matter a few moments' thought, and then pronounced: *The Age of Sharing*. Thank you, Aram.

This book would never have been written without the support of my family and friends. My parents, Rob and Judi, helped greatly at a crucial time, and my siblings, Zoë and Greg, cheered me on all the way. Mark Godfrey doesn't know quite how inspirational he has been as I have plugged away at this. S. will be thanked elsewhere.

I'm not much fun to be around while writing, so my deepest and most heartfelt gratitude goes to the people who live with me. Maya and Yasmin, who will mostly be excited simply to see their names on this page, have been waiting for this book for a while. We've had some great chats together about what sharing means to them, and their occasional inquiries into when I'm finishing the book motivated me more than almost anything else. But at the centre of it all stands Etty – her sacrifices and willingness to take the strain, alongside her encouragement, belief and support, enabled me not only to produce this text but quite simply make everything possible (the travel, the weekends, the deadlines...). Etty, thank you.

1
Introduction

In late 2015 an informal after-work event was held in Manhattan's Lower East Side for high-tech entrepreneurs in the sharing economy. One of the panelists was discussing different models of sharing. Some are based on sharing for free, she said, while others involve sharing for money. I waited for someone to raise their hand and challenge the speaker, but no one seemed put out by the sentence they had just heard. I had come across the idea of 'sharing for money' before. When I first encountered it I wondered whether it was not a simple contradiction in terms, like 'selling for free'. Or perhaps this was a classic example of ideology at work through language, where a word with positive associations is deployed in order to conceal the true exploitative order of things. But pointing to the misuse or even wilful abuse of the word 'sharing' is too easy and fails to contend with its shifting senses and multiple layers of meaning. The fact that someone could say 'sharing for money' and be understood by her audience, and the retort that the exchange of money negates the possibility of sharing, both require historical and cultural contextualization. The following pages are my efforts to do just that.

As a first step, we might observe how much sharing people seem to be doing these days. When we go online we share – photos, status updates, thoughts, memes, opinions, information. We are sharing offline too: witness the growth of the sharing economy. Powered by apps, people are sharing their spare rooms, cars, power drills, free time, expertise, couches, workspaces, dinner leftovers and pets. We are also sharing when we talk about our emotions, which we do more often and in more situations than any previous generation. Some people are taught how to communicate this way from a very young age: American preschoolers, for instance, sit in 'sharing circles', where they talk about their unique experiences while their classmates listen, awaiting their turn to share.

At the same time, the use of the word 'sharing' to describe some of these activities is contested. For example, certain critics of the sharing economy say that 'it isn't really sharing, it's renting/selling/trading'. Similarly, one might observe that Facebook does not share data with advertisers (though that is the language Facebook uses); rather, they sell it. Before engaging with these critical claims, though, I suggest that the very fact they are being made is indicative that something is at stake: the concept of sharing itself.

This is the age of sharing, then, because 'sharing' stands for both the cutting edge of our digital media-saturated capitalist society and economy, including the way we interact online, and a critical position vis-à-vis this society and economy. Sharing is both supportive and subversive of hegemonic (digital) culture: supportive in that the more you share updates and pictures on social media, for instance, the wealthier those platforms become, and subversive in that the more you share actual stuff with others, the less everyone needs to buy. Moreover, some say that sharing – be that of the distributive or communicative kind – leads to true and deep human connections.

In this book I will not be taking a stand on when the word 'sharing' is being used properly. In fact, my inquiries into sharing show that, as with many words, its 'proper' meanings and uses have changed quite drastically over time. For those who think that sharing is timeless, this discovery can be both surprising and perhaps a little destabilizing. In the following pages I show how the idea that sharing is the basis for authentic human relationships dates back no further than the 1930s, when city life, and especially advertising, were raising profound questions about authentic personhood. Moreover, the altruistic sense of sharing, or 'sharing as caring', only really

Figure 1.1   Sharing is caring; Cyanide and Happiness,
© Explosm.net, <http://explosm.net/comics/2432>

took root from the 1970s. By tracing changes in the meanings
of 'sharing' – and especially the entrenchment in the mid-
2000s of its sense as what we do online – this book shows
that the prevailing uses of the term today, and the criticisms
of these uses, have common roots in a sense of self moulded
by capitalism. Thus, while remaining agnostic as to the
'proper' way to use the word 'sharing', I am nonetheless alert
both to the role played by powerful social media companies
in disseminating one of its newer meanings, and to the interest these companies have in being associated with the concept's prosocial connotations.

Sharing is a very emotive concept: to start, it is deeply
associated with childhood, and learning to 'share nicely' is
one of the most basic skills preschoolers are expected to
assimilate; second, and relatedly, sharing is always good –
you cannot share non-nicely. Sharing, we are told, is caring,
and, as such, has a warm glow around it. This warm glow
also invites an ironic stance, as expressed in Figure 1.1:
calling something sharing (in pretty colours to boot) can
conceal its immorality; if we call it sharing, we might be able
to get away with anything. This cartoon thus neatly captures

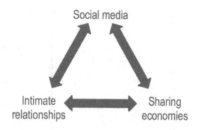

Figure 1.2 The spheres of sharing

a feature of sharing that is key to this book: it is both a practice or set of practices with ethical dimensions, and at the same time a word with ethical connotations. This book aims to explore them both.

Three spheres of sharing form the focal points of this book: sharing as the constitutive activity of social media; sharing as a model for economic behaviour; and sharing as a category of speech. In this way, the book has something to say about our technologically mediated social lives; about our economic lives as producers and consumers; and about our emotional, interpersonal lives. At first glance, these spheres are quite distinct, and there would not seem to be a prima facie reason for bringing them together. Is it enough that the word 'sharing' is associated with each of them? I argue that it is, because 'sharing' is an important part of how these practices are constructed and conceptualized; it is a metaphor in terms of which different spheres of sharing construct one another and themselves. This is represented graphically in Figure 1.2.

When we talk about sharing we implicitly or explicitly engage with a set of values. Later on, I shall demonstrate and elaborate on the ways that each of the three spheres of sharing discussed here enacts certain values. For now, suffice it to say that when we talk about sharing we are talking about purportedly prosocial behaviours that promote, or are claimed to promote, greater openness, trust and understanding between people. Hence, Mark Zuckerberg, founder and CEO of Facebook, can say that sharing (on Facebook) makes the world a more connected place; sharing economy evangelists promote sharing as a remedy for the ills of selfish and destructive hyper-capitalism (Botsman and Rogers, 2010); and Donal

Carbaugh defines the speech category of sharing as talk with a 'relational embrace' (Carbaugh, 1988).

It is the contention of this book that sharing, both as a broad category of social practices, and as the word used to describe a wide range of practices, is on the rise. Ours is the age of sharing.

What is Sharing?

There are many ways to answer this question. One is to inquire into the difference between sharing and other modes of resource management, such as buying (Belk, 2010) or lending (Larrimore, Li, Larrimore, Markowitz and Gorski, 2011). When answering the question this way we endeavour to isolate the characteristics of 'sharing-ness' in which acts of sharing partake. Then, equipped with these characteristics, we can decide whether or not to bestow the title of 'sharing' on different activities and argue with others over its aptness in different contexts. These arguments are played out in the field of file sharing, for instance, where we may hear comments such as 'It's not really sharing, it's online theft'. By talking of 'pseudo-sharing', Russell Belk engages in a somewhat similar strategy (Belk, 2014). But this is not what I mean when I ask what sharing is, and the objective of this book is not to demarcate its boundaries such that certain acts by definition fall beyond what may be considered as sharing. This is not to say that such an approach cannot be nuanced, and I recognize the complexities posed by borderline cases, but with that approach there is an intellectual push for a definition or conceptualization that can be used to categorize different practices and, as I have already intimated, that is not the direction in which I am pushing.

By contrast, the non-prescriptive approach adopted here starts by asking what kinds of things are called 'sharing' in practice. Indeed, my own interest in sharing began after noticing, at some point in 2010, that the word 'share' was all over the internet. This observation sparked my sociological imagination, and led me in search of other practices that are called 'sharing'. So if one way of understanding the question 'What is sharing?' is to see it as asking what falls within a predefined category and what does not, when I ask what sharing is I am asking which social actions and phenomena we actually call 'sharing'. This is the difference between asking 'What practices should we call sharing?', which is prescriptivist, and 'What practices do we call sharing?', which is what linguists would call a pragmatic approach.

There are a number of good reasons for preferring the latter to the former, and one reason for nonetheless keeping the former in mind throughout. The first and main reason for inquiring into the practices that we call 'sharing' (rather than asking whether use of that term is justified or not in different contexts) is that whatever else it is, 'sharing' is a metaphor, and when dealing with metaphorical usage the question of whether 'x really is y' is moot (the performer did not really bring the roof down; my erudite friend is not really a walking dictionary). The fact that diverse practices are termed 'sharing' should be taken as an opportunity to explore how various spheres of life are constructed through the use of metaphors from other spheres. One answer to the question 'What is sharing?' might thus be: sharing is a metaphor we live by (G. Lakoff and Johnson, 1980).

This is not to say that the literal, non-metaphorical meaning of sharing should not interest us. It should, because it will help us unpack the subsequent metaphorical uses as well as leading to insights as to what might be implied by the notion of sharing. The original meaning of 'sharing' is given by the *Oxford English Dictionary* as dividing, or splitting. When understood this way, the linguistic proximity between 'shearing' and 'sharing' is suddenly obvious. Similarly, we thus realize that a 'ploughshare' is so called not because it was shared by all of the villagers, but because it shared, or rent asunder, the earth.[1] This sense of division is central to the early meaning of sharing, and also to our naive understanding of the concept: sharing is when you let others have some of what's yours. Sharing is thus about division and distribution, and as such raises questions about distributive justice. What, we frequently ask, is a fair share? This is an important question, because when sharing is about dividing and distributing resources, it is a zero-sum game; there's only so much to go round. If sharing is conceptualized as an ethical practice, it is because of its original relation to the distribution of material resources.

As mentioned, today there is a wide range of practices that are referred to as 'sharing', including posting updates on social network sites, or more generally the digital transfer of information; certain ways of exchanging goods and services; and talking about our emotions, or conveying information, usually verbally, of affective import. Given this, perhaps one could extrapolate a set of core values that inhere in the practices we call 'sharing'. On the other hand, perhaps the polysemy of the word 'sharing' today is no more than homonymic: we use the same word but really it has quite different meanings in different contexts. One of the objectives of this book is to show that the different senses of sharing today are related. At the very least, when a practice is called 'sharing' a certain stance between the participants in that practice is posited; this stance might involve values such as openness, trust and maybe a sense of commonality.

However, because we are reflexive social actors, a second-order problem arises: we know which values are enacted when we talk about sharing (hence the cartoon in Figure 1.1), and so we may then want to describe certain practices as sharing in order to associate these values with those practices. Here, the word 'sharing' takes on a rhetorical force for the sake of which it is deployed or, indeed, avoided. For instance, Robin Chase, founder of Zipcar, explained how, during the company's early days, she forbade use of the term 'car-sharing', believing it to have negative connotations (Levine, 2009). Today, though, the term appears prominently on Zipcar's 'About' webpage (four times in the first four sentences, to be precise).[2] As already noted, the rhetorical force of calling a practice 'sharing' has been observed by critics of file sharing, who disapprove both of the practice and the name given to it. Moreover, bearing in mind the positive values usually associated with 'sharing', there is something jarring about reading the privacy policies of the Facebooks and Googles of the mediascape and learning what information they do or do not 'share' with advertisers or law-enforcement authorities. At this point we would seem to be swinging back towards the view that there are certain practices that are 'not really sharing': social network sites (SNSs) do not share information with advertisers, we might want to say, they sell it, and this distinction would seem to be absolutely crucial to any understanding of the political economy of the internet and social media. This example encourages us to inquire into the usage of the word 'sharing' in this specific context (see Chapter 3, and also the case of 'file sharing' in Chapter 6), which leads us to the quite morally neutral deployment of the term in the context of computing from its earliest days in the 1950s, when time sharing was the mechanism for enabling access to computers by as many users as possible.[3]

It is my contention, then, that through an analysis of 'sharing' we gain insights into contemporary culture, and especially contemporary digital culture. In this regard, 'sharing' might be considered a keyword for the digital age (John, 2016). In Raymond Williams' canonical *Keywords* (1976), he showed special interest in 'the explicit but as often implicit connections which people were making, in what seemed to me, again and again, particular formations of meaning – ways not only of discussing but at another level of seeing many of our central experiences' (Williams, 1976: 15). This book examines these 'implicit connections' and pays particular attention to the word 'sharing' precisely because it pertains to 'many of our central experiences' today: our lives online; our lives as economic beings; and our lives as lived through our interpersonal relationships.[4]

Keywords tell us important things about the culture in which they operate. They do not encapsulate the entire culture – no one word or symbol could do that – but their analysis sheds light on enough aspects to make the effort worthwhile. When I call 'sharing' a keyword, and thereby make a claim as to the importance of paying attention to it, I mean that if we study 'sharing' and the social spheres in which it is a significant concept, we are able to learn something about how those spheres are related. But more than that, positing an age of sharing implies that this word, this metaphor, is itself fundamental to the ways those spheres are related.

So far I have pointed to the spheres of sharing that this book focuses on, and I have noted that the original, literal meaning of sharing is about the physical division of material resources. Before pressing on, I would like now to say more about the meanings of the word 'sharing'; conceptions of sharing as the fundamental state of both individual people and humanity as a whole; sharing as straddling the nebulous public/private divide; and, in closing this introduction, something about the chapters that lie ahead.

The Meanings of Sharing

Although the belief that a dictionary will provide us with the meaning of a word may be misguided (Williams, 1976), it is sometimes a good place to start nonetheless. Starting with the dictionary definition of 'sharing' has already taught us that it referred – and still does refer – to the division, or distribution, of resources. Sharing as distribution is, of course, governed by cultural norms. These norms are the subject of Katriel's description of ritualized sharing and exchange among children (Katriel, 1987, 1988), as well as constituting one of the main focuses of the early anthropology of hunter-gatherer societies (see, for instance, Morgan, 1881; Stefansson, 1913). From these studies it is clear that sharing, whether it involves the distribution of either candies or prey, is constitutive of social relations. Indeed, Katriel, drawing on Mauss (1966 [1925]) and others, views the sharing of treats among children as 'a ritualized gesture that functions to express and regulate social relationships with the peer group' (Katriel, 1987: 307).

Another meaning of sharing is to have something in common with someone, where this thing may be concrete or abstract. For instance, when students share a dorm room, the room belongs to each person, and itself remains whole, despite being shared. This logic also applies to abstract shared objects which cannot be owned, such as interests, fate, beliefs or culture. Here too sharing is about distribution, but in an abstract and passive way, and in a way that is not a zero-sum game: the fact that a belief, for instance, is shared by two people does not preclude other people from coming to share that belief as well. Significantly, with this type of usage of 'sharing', there is no 'sharer'. We share this planet on which we live, but no one is actively sharing something with others; rather, the sense of sharing in this context is closer to 'partaking of'. Nonetheless, this sense of sharing still implies social bonds: people who share a fate are bound together by that fate; people who share a belief in a certain deity have that in common; and, some say, the people who share this planet are thus obliged to it and to one another.

In addition to being an act of distribution, sharing can also be an act of communication. This is the case when we talk about sharing our feelings or emotions. Unlike the two previous meanings of sharing, which the *OED* dates to the sixteenth century, this sense of sharing, as imparting one's inner state to others, would appear to be somewhat newer, but it has quickly become a central social practice. Indeed, the first citation provided by the *OED* for the meaning of 'sharing' as 'to impart to others one's spiritual experiences' only dates back to 1932 and is offered in the context of the Oxford Group, a Christian movement popular in the 1920s and 1930s (of which more in Chapters 2 and 5). In this regard, the *OED* quotes A. J. Russell, who wrote that the Oxford Group defined 'sharing' as meaning 'Confession and Witness' (Russell, 1932). To be sure, the Oxford Group attributed great importance to the practice of sharing: as described by Dick B. in his hefty volume on the spiritual roots of Alcoholics Anonymous (where, as with all support groups, sharing is the constitutive activity), 'Almost every Oxford Group book abounds with discussion of "Sharing" by confession. The emphasis was on sharing with God and with another' (B., 1997: 326). From here, it is but a short step to the notion of sharing one's feelings that is central to the formation and maintenance of intimate relations in contemporary western society.

Sharing and the Human Condition

The last few years have seen a glut of books in praise of humankind and its tendency for sharing, cooperation and generosity. At the same time, a psychological literature has emerged that looks at propensities to kindness and sharing at the individual level. My objective here is not to evaluate the scientific validity of this work, but rather to point to the

Figure 1.3   ecoSharing.net

narrative that it represents, which somewhat tritely can be put as follows: we (both humanity as a whole and individual human beings) have a natural tendency to share which is beaten out of us over time by the vagaries of capitalist culture. (Another version of this argument is that 'new' technologies, such as the internet in the 1990s and SNSs in the 2010s, are making us more cooperative. I shall come back to this point in my discussion of the construction of the sharing economy as a technological phenomenon in Chapter 4.)

At the individual level, pro-sharing texts present work carried out by developmental and evolutionary psychologists that show children to have sharing and cooperation hard-wired into their behaviour. For instance, a darling of the collaborative consumption community is Michael Tomasello, whose *Why We Cooperate* (Tomasello, 2009) is referenced by Rachel Botsman and Roo Rogers in their book *What's Mine is Yours* (2010), and is included in Shareable.net's fifteen best books for 2009. The message that Botsman and Rogers and the Shareable.net website take from the book is that 'preverbal children have a natural inclination to share and help others'[5] and that '[c]hildren are sociable and cooperative by nature' (Botsman and Rogers, 2010: 69). This can also account for the use of children in images on websites that enable people to share stuff they own. For instance, the now-defunct ecoSharing.net used a picture of two smiling girls, around six years old, in a park listening to an iPod together with one earphone each, above the text, 'Do you remember how much you enjoyed sharing what you owned?' (see Figure 1.3).[6] By adopting practices of sharing – here understood as sharing stuff we own – we are returning to an earlier, more innocent and simple state of being.

The purported naturalness of sharing is also conveyed through representations of premodern, or even prehistoric, human societies, which are perceived as living in a more 'natural' way than our current lifestyle. For instance, on the 'What we believe' page of the On the Commons movement's website, it is claimed that '[t]o work on the commons is to work to enliven the deep and ancient memory we all hold of egalitarian and reciprocal relationship, of belonging, of authentic community, and of love, wonder, and respect for the natural world'.[7]

It is important to reiterate that I am not commenting here on the accuracy of this reading of the human psyche or of the development of humankind, but rather making the point that this is the narrative very often evoked in analyses of sharing. What makes the narrative so powerful is that it operates at both the individual and societal level: both individual humans and society as a whole start off cooperating and sharing; with time, though, they are corrupted by capitalism and hyper-consumerism. A return to sharing, according to its proponents, is thus both a return to a more natural state of being and morally superior. These are issues to which we shall return in greater depth in Chapter 4.

Research into 'Sharing'

This is not the first text to have been written on sharing. Nor, I am sure, will it be the last, as interest in sharing appears to be sharply rising, particularly in relation to online sharing and sharing as a consumer behaviour. Special issues of journals are being published on sharing,[8] and conferences and study days are being held.[9] There is a notable surge of interest in the academic community – and beyond – in the notion and practices of sharing.

Of particular interest are texts that explicitly engage with the concept of sharing itself. Thus, for instance, while fascinating, Bart Cammaerts' article on 'disruptive sharing' (Cammaerts, 2011) and David Brake's (2014) and Graham

Meikle's (2016) books on social media sharing teach us more about how the word 'sharing' is used in its 'natural environment' than about the contested meanings of the word itself and the role it plays in constituting digital culture; likewise, Alfred Hermida's bestseller, *Tell Everyone* (2014). The same is equally true of most of the anthropological literature about food sharing among hunter-gatherers (though for a notable exception, see Bird-David, 2005).

Some previous studies of sharing, however, do engage explicitly with sharing. Without a doubt, the influential writings of Russell Belk have given an enormous impetus to the study of sharing. Belk approaches sharing as 'a fundamental consumer behavior' and seeks to establish it as a category distinct from commodity exchange and gift-giving (Belk, 2010: 715). It is, he says, a 'third form of distribution' (Belk, 2007: 128). This approach also leads him to coin the neologism 'pseudo-sharing', referring to 'a business relationship masquerading as sharing' (Belk, 2014: 11; see also Eckhardt and Bardhi, 2015). In Chapter 4 on the sharing economy, we shall discuss at length the idea that there are practices called sharing that 'are not really' sharing. For now, let me just put on the record my divergence from this perspective. This is not because I think that lending your neighbour your drill is the same as renting your neighbour your drill, with a small cut going to the internet company that mediated the transaction. However, when the notion of the 'sharing economy' includes both of these practices, I propose that we stop to think critically about what this means in terms of sharing, rather than adjudicating which practices are 'true sharing' (Belk, 2014) and which are not. And, in any case, there is a similarity in that for both transactions a private possession of yours is handed over to a stranger for a limited period of time, suggesting there is more than just a 'sematic confusion' involved. Be that as it may, Belk's work, which is remarkable for its breadth, is clearly important, and has inspired a tranche of studies of sharing as a form of consumption (for a smattering, see Albinsson and Yasanthi Perera, 2012; Bardhi and Eckhardt, 2012; Harvey, Smith and Golightly, 2014; McArthur, 2015; Ozanne and Ballantine, 2010).

Sharing has also grabbed the attention of media and communication scholars. An early contribution in this regard was made by Felix Stalder and Wolfgang Sützl, whose special journal issue on the ethics of sharing was extremely timely. In their introduction, they start by declaring that 'Sharing has emerged as one of the core cultural values native to the networked environment' (Stalder and Sützl, 2011: 2). And Andreas Wittel (2011), for instance, has engaged with the 'qualities of sharing and their transformations in the digital age'. The grist for Wittel's mill is the economic approach to sharing represented by Yochai Benkler (2006) and digital enthusiasts à la Leadbeater, Shirky, and Tapscott and Williams (Leadbeater, 2008; Shirky, 2010; Tapscott and Williams, 2006). In contrast to them, he looks at the 'social qualities of sharing', and the impact of digital technologies on them (Wittel, 2011: 5). Wittel makes the useful distinction between material and immaterial objects of sharing, noting, as I did above, that sharing immaterial objects entails no sacrifice. He also observes that the digital age is effecting changes to 'the notion of sharing itself' (Wittel, 2011: 6). However, like Belk and others, he also suggests that 'the term sharing is rather problematic, perhaps misleading, for digital objects'. He adds, 'It seems that sharing, like stealing, has entered the language of digital cultures due to mere ideological reasons' (p. 6). Like Wittel, I too am interested in the ideologies surrounding 'sharing', but for me its use is not 'problematic'. Viewing 'sharing' as a metaphor, and in line with the broadly pragmatic approach to word usage adopted throughout this book, the application of the term 'sharing' to new practices raises questions about online sociability, digitally mediated forms of communication, and conceptualizations of property that a close analysis of 'sharing' can help answer.

Another important contribution has been made by Jenny Kennedy. Like Belk and Wittel, she too is alert to the 'semantic richness' of sharing and observes that social media platforms are able to exploit 'its association to predated activities' in order to promote our use of them (Kennedy, 2013: 129; see also Kennedy, 2016). But rather than prescribing our use of the term by pointing to practices that 'are not really' sharing, Kennedy observes instead that the 'ubiquity and everydayness of the term sharing belies the diverse and complex social, cultural, economic, and political processes it is employed to describe' (p. 135). I am entirely sympathetic to this stance. However, I go one step further, suggesting that the complexity of these processes inheres in the very term 'sharing' itself. One might argue this is what Raymond Williams meant when he said that 'some important social and historical processes occur within language' (Williams, 1976: 22).

Of current theorists of sharing, Kennedy's approach is the closest to mine. To start, unlike most commentators, she explicitly acknowledges that sharing is 'a distinct form of communication' and pays special attention to networked culture (Kennedy, 2014: i). She unpacks three main 'discursive threads' around sharing which are close – though not identical – to my three spheres of sharing (Kennedy, 2016: 5). She refers to 'sharing as an economy', 'sharing as scaled distribution' and 'sharing as social intensity', where she defines sharing in relation to 'disclosure and affect' (p. 468). My own focus on the 'sharing economy', on online sharing, and on sharing as the type of communication that sustains our therapeutic culture, clearly resonates with Kennedy's work. Furthermore, following Nick Couldry (2012), Kennedy adopts a practice approach, asking what 'people [are] doing in relation to media across a whole range of situations and contexts' (Couldry, 2012: 39; cited in Kennedy, 2016: 469). Indeed, the epigraph for Couldry's chapter, 'Media as Practice', reads: 'What media are needs to be interrogated, not presumed' (Larkin, 2008: 3; cited in Couldry, 2012: 33). Kennedy seems to take this as applying to sharing; I would wholeheartedly concur.

Additional texts about sharing will be encountered in the following pages. The scholars I have just mentioned, though, are particularly useful for me. Belk is a beacon, illuminating the paths we might wish to take, even if I do not accompany him all the way down the one he pursues; Wittel is among the first group of researchers to pay critical attention to the specifically digital qualities of sharing; and Kennedy's practice-theory approach to sharing is remarkably fruitful and sits extremely well with the non-prescriptivist approach to sharing presented here.

What this book adds to these discussions of sharing is not just its communicative aspect, but the insight that sharing is a type of communication that implies a certain style of interpersonal relationship, one that is based on honesty, openness, mutuality, caring, equality, trust and fairness. These are the declared values of the sharing economy as well (see, for instance, Buczynski, 2013), and also form a part of the internet imaginary (Flichy, 2007). I already hinted at these values earlier, and they shall return throughout the book.

The Rest of the Book

The values just mentioned are part of the promise of sharing today. However, 'sharing' has not always been associated with them. In Chapter 2, I ask when sharing became caring and undertake a diachronic analysis of 'sharing'. The analysis is based on around 4,500 instances of the word 'sharing' in English-language texts spanning over two centuries and enables me to locate the gradual entrance into 'sharing' of the values associated with it today.

With Chapter 3, I start to engage with the first of the three spheres of sharing and discuss sharing in relation to the internet and social media. My objective here is not to ask why we share and what we share, but rather to critically examine the prominence of the word 'sharing' in this context. For instance, I show how the word 'sharing' has been applied retroactively to the early internet. Scholars and other commentators have written about the prosociality of the internet; some have said that the internet is, and has always been, about sharing. However, texts about the internet from the 1990s and early 2000s do not actually refer to 'sharing', suggesting that the explicit affiliation between sharing and the internet was made more recently. An analysis of the homepages of forty-four SNSs between the years 2000 and 2010 confirms this, showing the years 2005–7 as the time when 'sharing' became the *sine qua non* for online participation, pointing to the centrality of powerful media organizations in pushing the word to its current prominence. One exception to this is the field of hacking, where sharing has always been talked of as being a key value and practice. This, of course, reinforces the idea that sharing has specifically digital connotations today, a point that will recur throughout.

Chapter 4 is about sharing economies. As with online sharing, the focus here is on the work that the word 'sharing' does in this context. By talking about sharing economies I cast my net wider than what is known today as 'the sharing economy'. In particular, the chapter discusses sharing economies of production – as instantiated by Wikipedia and open source software, for instance – which is where the term was used before it was applied to acts of consumption. Drawing on an analysis of newspaper articles about collaborative consumption, I show how this part of the sharing economy has been discursively constructed as technological and digital. This is the field where we are perhaps most likely to hear people saying that 'it isn't really sharing' on the grounds that money is often involved. However, I do not add my voice to that particular chorus, preferring instead to inspect the argument quite closely (no one says, for instance, that shares (as in 'stocks and...') should not be termed thus because they involve money), and asking whether the adoption of the 'sharing' metaphor in this regard nonetheless teaches us something.

Following that, Chapter 5 offers an exploration of sharing as a category of talk. Here, I locate the emergence of sharing as a type of talk in the public confessional practices of the Oxford Group in the 1930s. The Oxford Group was a Christian movement that practised the confessing of sins in a group setting, terming this practice 'sharing'. The Oxford Group was the forerunner of Alcoholics Anonymous and as such, I argue, holds an important place in the emergence of our contemporary therapeutic culture. In this chapter, I track 'sharing' from the Oxford Group through reality TV to social media. In all of these cultural locales we find similar (but not necessarily identical) assumptions about the modern self, and especially assumptions concerning authenticity and the value of making our inner selves public, or at the very least known to another person. There is an important sense, then, in which sharing, as a type of communication, is constitutive of our intimate (and other) relationships in that it reflects and constructs expectations for honest and authentic communication between equals.

Chapter 6 is an exercise in implementing what the previous chapters tell us about sharing. The object of the exercise is file sharing. The chapter opens with a discussion of the term 'file sharing' and the debates surrounding it ('it's not really sharing' makes another unsurprising appearance). This discussion is supplemented with an analysis of over 450 posts to file-sharing forum debates about the ethics of a certain form of file sharing known as 'torrenting'. I show how the presence of the metaphor of 'sharing' in the debates shapes some of the positions adopted in it. Or, put differently, I show how, for at least some of the file sharers engaged in interactions over the best (most ethical, most efficient) way to share files, 'sharing' really is a metaphor they live by.

Two Notes about Style

As the reader will no doubt already have noticed, the word 'sharing' sometimes appears in quotation marks, and sometimes it does not. Stefano Predelli (2003) discusses quotation marks, observing that they can serve to signal distance from the term enclosed within them. Sometimes, then, when people put 'sharing' in quotation marks what they mean is that what they are talking about is not actually sharing; in fact, they even suggest a degree of hostility and convey that although they are using this word, they do not think it is the appropriate one. In my reading about sharing, this is a use of quotation marks that I have come across very many times, and conveys a similar meaning to air quotes (the kind we make in the air with our fingers). However, when I put 'sharing' in quotation marks, it is usually to signify that I am talking about the word or term 'sharing', rather than practices of sharing. This explains why I can say that this book is about sharing and 'sharing'. However, I do not promise uncompromising consistency, and I trust the reader will understand whether the discussion is about sharing or 'sharing' from the context. If in doubt, it is probably 'sharing'.

The reader may also notice the use of the first-person plural ('our society is...', 'we understand that...' and the like). This 'we' is not intended to fetishize the experiences of people living in modern, western, media-saturated societies, though it does assume a certain commonality between them.

Where the use of 'we' may be exclusionary is in relation to non-English speakers. In particular, some languages have a different word for 'sharing' depending on whether what is being shared is 'your candy bar' or 'your emotions'. In other words, some languages have different words for the distributive and communicative logics of sharing. The coming together of these two types of meaning under a single word has implications for how the word is understood by English speakers. This is a crucial part of the arguments made throughout this book. Readers familiar with languages other than English will know better than I whether the arguments made here about 'sharing' apply equally to *teilen*, *partager*, *compartir* and more.

2
How Sharing Became Caring

This chapter inquires into the differences between past and present meanings of 'sharing', the cultural shifts that they are associated with and how they might be charted. Given that 'sharing' today has strong normative connotations, this chapter can be read as an attempt to understand how and when sharing became caring.

A key insight on which this book rests is that the particular configuration of meanings that constitute the word 'sharing' today make it both a term with special power in contemporary culture, and a term through which to understand that culture. It is able to do the cultural work described in this book precisely because 'sharing' is a word that enfolds within it a group of meanings that both reflect and reproduce important aspects of contemporary society. That is, when we – English-speakers in western societies – hear talk about sharing today, we understand the concept differently from both our grandparents and their grandparents (for these purposes it does not matter whether you are 16 or 116). Obviously, the notion that sharing might refer to an act of computer-mediated communication would have been foreign to inhabitants of the 1970s, and even the 1980s, while the idea that intimate relationships are based on sharing would not have been part of people's mindset in the 1950s. On the other hand, conceptions of sharing as a mode of distribution would have been common not only to our great-great-grandparents but to their great-great-grandparents as well. Moreover, as we shall see, the various meanings of sharing today have traceable roots in earlier senses and uses of the word.

What I present here is a pragmatic analysis of 'sharing'. Jef Verschueren defines pragmatics as 'the study of linguistic phenomena from the point of view of their usage properties and processes'; it is, in other words, 'the study of language use' (Verschueren, 1999: 1). According to this approach, meaning is 'dynamically generated in the process of using language' (p. 11). Furthermore, while this book as a whole adopts a synchronic view of 'sharing', seeking to understand the interplay of its range of meanings and connotations in the present (de Saussure, 2011 [1916]), this chapter takes a diachronic approach, examining how and when 'sharing' came to have the meanings that it does today.

'Sharing' in the Dictionary

The first entries for 'sharing' or 'share' in the *Oxford English Dictionary* date to the mid-sixteenth century and have the meaning of cutting into parts, or cutting off. However, even this very physical sense of sharing may be metaphorical, as perhaps the earliest use of the word – the Old English 'scearu' – was to refer to the groin, where the trunk of the body divides into two legs.[1] 'Sharing' in the sixteenth century was close in meaning to 'shearing'. Examples of this from the *OED* include: the 'cultor shares the soyle', and 'Aples [are] shared in peeces'. More generally, we learn that 'sharinges' are akin to 'fragmentes'. This sense of division was also applied to more abstract entities than apples, such as kingdoms, countries and wealth.[2] Here, the verb 'to share' implies someone who is doing the sharing, dividing things up. The *OED* also points to the meaning of sharing as having something in common with others. This might be a fate, expenses, the 'love of hard fighting' or a sickness.

In the *OED*'s quotations for 'share' and 'sharing', the word is quite neutral. There appears to be no moral aspect to the division and allocation of possessions referred to by the word 'sharing', even through to the nineteenth century, and instances of people sharing a certain condition are presented very matter-of-factly. In other words, there is nothing in the *OED* that suggests that sharing was good, or something to be strived for, certainly not in the way that it is seen today. We can see this in a quotation from 1860, taken from a legal tome written by Harvard University law professor, Emory Washburn. The quotation concerns a particular arrangement between agricultural landowners and tenants, and refers to 'sharing' as a way by which the latter might pay the former: 'There is a mode of letting lands', writes Washburn, 'not unusual in the country, where the tenant is to cultivate them, and then share the crops with his landlord' (Washburn, 1860: 364). Unfortunately, we do not know whether contemporary readers of this text protested this use of the word 'share', stating that this is not really sharing, but rather paying rent.[3] However, its inclusion in the *OED* under the definition 'To grant to give another or others a share in' suggests that this would have been quite an acceptable way to use the word. Moreover, we might continue, the unequal power relations between landowner and tenant make it absurd to talk about the tenant sharing his crops with the landowner. But this would be to anachronistically bring our understandings of and associations with 'sharing' to quite a different cultural context. In fact, by doing so, we would be revealing quite a lot about our assumptions about sharing today, for instance, that 'sharing' and 'paying' are incompatible; that sharing takes place between peers and implies a flattening of hierarchies; and that sharing is a voluntary act. Today, we would no more readily describe the relationship between landowner and tenant as one of sharing as we would say that the child shared her candy bar with the class bully under threat of a beating. Today, 'sharing' implies equality, mutuality, a relationship freely entered into. The question to be answered is, when did these values become associated with sharing?

At least part of the answer to this question lies in one of the newer meanings of sharing that the *OED* presents. In addition to the meanings of sharing as dividing and having in common, it also has the newer sense of 'to impart to others one's spiritual experiences'. The *OED* relates this meaning quite specifically to the Moral Rearmament movement and the Oxford Group, a religious movement whose distinguishing practice was a kind of public confession of one's sins in the presence of other group members, a form of communication they termed 'sharing'. We shall see in Chapter 5 how sharing is constitutive of today's therapeutic culture and I shall examine in some detail the place of the Oxford Group in establishing this sense of sharing. For now, though, it is enough to note that the *OED* puts the solidification of the meaning of *sharing-as-telling* in the 1930s. In this sense, sharing was said to bring a kind of redemption to the sharer, and to enhance a sense of fellowship among those present. Today, this kind of therapeutic sharing is readily associated with Alcoholics Anonymous and support groups in general: in such settings the participants' sharing is meant to help them overcome a difficulty, and it also creates a sense of togetherness among the group. This is a key moment in sharing becoming virtuous. Significantly, Alcoholics Anonymous was actually an offshoot of the Oxford Group. In the Oxford Group, and later in the support group, sharing-as-telling comes to signify a level of solidarity, equality and the absence of hierarchy, trust, openness, the expression of one's true (flawed) self, and reciprocity. With the emergence and solidification of this sense of sharing, a new semantic field was brought into being.

So far we have adhered to the *OED*'s definitions of sharing; let us now turn to a much more detailed account of the shifting meanings of sharing and its affiliation with a set of prosocial values.

'Sharing' and Corpus Analysis

Raymond Williams warns us against relying too heavily on dictionaries for our understanding of words, 'and especially for those that involve ideas and values' (Williams, 1976: 17). Indeed, when the word we are interested in is a key word (Ortner, 1973; Wierzbicka, 1997), or a 'cultural concept that is dense' with meaning (Carbaugh, 2007: 177), then clearly the dictionary will not suffice. Moreover, as noted by linguist Deborah Cameron, 'Even the most conscientious lexicography never gives more than a partial indication of the range of use of words at any given time because of the inevitable limitations of its source material' (Cameron, 1995: 126). Thus, while the *OED* does define 'file sharing', for instance, it does not define 'sharing' as referring to computer-mediated communication; nor does it show the ways in which 'sharing' has come to stand for all that is good in romantic and other intimate relationships. In order to offer a more detailed account of the evolution of 'sharing' we certainly do need to go further than the dictionary can take us. To this end, I present a (mostly) qualitative corpus analysis. Qualitative corpus analysis is 'a methodology for pursuing in-depth investigations of linguistic phenomena, as grounded in the context of authentic, communicative situations that are digitally stored as language corpora and made available for access, retrieval, and analysis via computer' (Hasko, 2012: 4758). In this instance, it means looking at naturally occurring language to identify changes in a word's meaning.

The corpora used here have been produced on the one hand by an enormous multinational corporation – Google – and on the other by a university-based sociolinguist – Mark Davies of Brigham Young University. Google's corpus has been produced through its Google Books project. In 2010, the corpus included hundreds of billions of English words from over 15 million books (Michel et al., 2011). However, Google's Ngram interface is relatively limited, and in this regard the corpora produced by Davies are far more useful. Here I shall be making use of two of Davies' corpora: the *Corpus of Historical American English* (COHA), which contains over 400 million words and ranges from 1820 to 2009;[4] and the *Corpus of Contemporary American English* (COCA), which contains 450 million words and ranges from 1990 to 2012.[5] Most of what follows is based on an analysis both of the 4,417 instances of the word 'sharing' from the COHA set and of collocate queries aimed at finding the words that appear in proximity to 'sharing' most often.

The first observation to extract from both Google's data and COHA is that talk of sharing has increased greatly since the early twentieth century, and especially since the 1970s (see Figure 2.1). Although the two lines do not overlap perfectly, pointing to differences between the corpora, they

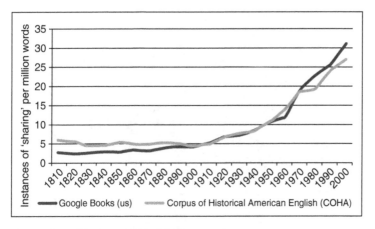

Figure 2.1   Sharing, 1810–2010

certainly overlap enough to enable us confidently to assert that we talk and write about sharing far more now than we did in the recent and not-so-recent past.

In addition to this quantitative change in talk of 'sharing' there has also been a qualitative change, which we can observe by making use of one of the corpus analyst's favourite concepts – collocation. Collocation involves identifying and counting the words that appear within a certain distance of what is called the *node word* (in this case 'sharing'). The assumption here is that the words that appear along with the node word can give a sense of the context in which that word is used. By drawing on a historical corpus, it is possible to pinpoint with relative accuracy the emergence of new senses of a word. To offer one obvious example, in the nineteenth century the words 'file' and 'sharing' do not appear in proximity to one another. This somewhat mundane observation is rendered visually in Figure 2.2, in which the darker the block, the higher the incidence of the word on the left-hand side in proximity to the word 'sharing'.

Some of the collocates here are unsurprising and not especially illuminating – such as that between 'sharing' and 'file'. What is interesting, though, is the ascendance of the meaning of 'sharing' that I have termed sharing-as-telling. For instance, prior to the 1980s, information was not something

Figure 2.2 Collocation heat map for 'Sharing', 1820–2010; the darker the block, the greater the collocation of the word with 'sharing'

that was 'shared'. Based on *COHA*, the main collocates for 'information' for the first half of the nineteenth century were words such as 'procured', 'afforded' and 'furnished'. Information, and indeed knowledge, only really became something that was shared from the 1970s. Nor were experiences or feelings 'shared' in the nineteenth century; as I shall show below, when they did appear along with the word 'sharing' it was usually in the sense of 'having in common with', or 'partaking in', and not in the word's active sense of telling another about one's experiences or feelings. Some collocates are not especially new: talk about profit sharing, for instance, has been prevalent since the 1870s, and it is hard to detect a pattern in relation to the collocation of 'sharing' and 'joy'. 'Fate' is a noteworthy entrance in this list, as the heat map shown in Figure 2.2 would suggest there is much less talk about shared fates now than in the mid-nineteenth century.

It is important to note, though, that this is not merely an exercise in counting, and Figure 2.2, while illustrative, tells only part of the story. In the following, I am interested in teasing out new ways that the word 'sharing' has come to be used. A new usage will be taken to represent a cultural shift of some kind. Sometimes this is obvious: of course English speakers in the US in the nineteenth century did not talk of 'file sharing'. However, it is less obvious, and perhaps eye-opening, that they did not talk about sharing secrets either: in the nineteenth century, secrets were told, revealed or discovered, but not shared.

The Emergence of Sharing-as-Telling

A crucial change in 'sharing' is that it has taken on a *communicative* sense, alongside its senses of *distribution* or *division* and *partaking in*. That is, it has joined the set of 'cultural terms for communicative action', defined as 'terms and phrases that are used prominently and routinely by people to characterize communication practices that are significant and important to them' (Carbaugh, Berry and Nurmikari-Berry, 2006: 205). Indeed, it is one of Donal Carbaugh's 'Fifty Terms for Talk' (Carbaugh, 1989).

As a term for talk, 'sharing' has increasingly come to designate the imparting of an internal state to others, and to represent an ideal in intimate relationships. The metaphor has shifted: in the phrase 'a problem shared is a problem halved', for instance, the metaphor is very close to the original sense of sharing as dividing. If I share a problem with you, then it is halved, just as if I were to share a candy bar with you; having halved the problem, it now weighs less heavily on me. Unlike sharing a candy bar, however, the sharer is not making any kind of sacrifice; indeed, the person being shared with is the one carrying out the more valuable work here. Be that as it may, it is easy to see how in this instance 'sharing a problem' can come to refer to an act of communication rather than of (metaphorical) division, and while precursors of this can be seen in the corpus from the nineteenth century, this sense of sharing as an act of communication is actually relatively new.

In fact, in the early instances in *COHA* of 'sharing' being used in relation to emotions or feelings, it is not in today's sense of 'telling', but rather in the sense of 'having in common'. For example, in a novel published in 1824, a character describes watching his brother die on the battlefield thus: 'Strange as it may seem, I who could never, from my infancy, see him suffer pain without sharing in it, took the cartridges from his quivering hand' (Sigourney and Lathrop, 1824: 191). A similar usage is found in *Moby Dick*, when two sailors, Stubb and Flask, are said to be 'sharing his [Star-buck's] feelings', meaning that they have similar feelings to Starbuck (Melville, 1851). A move towards the idea that sharing can be telling might perhaps be elicited from a scene in a children's book, also published in 1851, in which a little girl tells her brother about how she had been teased. After their conversation, we read that, 'When they went home, Fanny was as happy as ever again, for she found that her heart was very much lightened by sharing her troubles with her brother' (Jessup Moore, 1851: 154). Here, though, it is not that Fanny feels better for having *talked* with her brother per se, but rather that she has made her troubles lighter by sharing – that is, *dividing* – them. And, indeed, when encouraging Fanny to tell him what is making her sad, Frank, the brother, does not tell her that talking about it will make her feel better – a central component of the twentieth-century 'therapeutic gospel' (Moskowitz, 2001) – but rather that 'he should be very unhappy if he did not know the cause' (Jessup Moore, 1851: 153). In this case, as opposed to later uses of the word, Fanny's troubles are shared because she talked about them; their being shared is a consequence of having talked about them. The talking itself is not called 'sharing'; rather the talking leads to the sharing – or dividing and redistribution – of her troubles.

Likewise, in *Little Women*, Louisa Alcott has a character say: 'There is no rest for me until I have tried to lighten this burden by sharing it with you. Let me talk and wear myself out, then you shall help and comfort me' (Alcott, 1873: 111). We can see the relationship between sharing a burden and talking, but once more the relationship is that the former is achieved by means of the latter, and not that the two are coterminous, which they become only later on. Moreover, there is no suggestion that the act of talking is in itself helpful and comforting: first Louisa will talk, then she will be comforted. Nonetheless, although the metaphor of sharing is still related to the physical act of division, it is also worth noting that here, as in the previous example, the ability of talk to bring two people closer together is recognized.

The deepening of this discursive connection between 'sharing' and the inner emotional states of others can be seen in an entry from the corpus from 1899, where music critic Henry Theophilus Finck holds forth on the nature of sympathy – a capacity that characterizes the 'civilized man', but not the 'lower strata of humanity': 'Sympathy means sharing the pains and pleasures of another – feeling the other's joys and sorrows as if they were our own, and therefore an eagerness to diminish the other's pains and increase the pleasures' (Finck, 1899: 400–1). Today we would probably call this trait 'empathy', but the point stands: sharing does not refer to the act of telling another about one's 'joys and sorrows', but rather to the condition of sympathizing (or empathizing) with them. In the nineteenth century, then, talk of sharing in relation to emotions refers not to the person experiencing the emotions, but to an other who, by identifying with the emotions, is thereby said to be sharing (in) them.

In other words, at the turn of the twentieth century, the word 'sharing' had not yet come to take on its current meaning as a type of speech, and even when one shared by means of talk, it was still in the sense of dividing one's troubles or burden. This metaphor of sharing – the physical one – is also familiar from religious talk, wherein people unburden themselves of guilt by confessing their sins. Guilt weighs us down; sharing that weight – giving some of it to another – helps us, just as sharing a physical load lessens the strain on our back: thus, in Exodus (34:6–7), we learn that if we tell God our iniquities he will bear them himself. Moreover, it can also restore a relationship to a previous, better state. The Catholic confession, for instance, mends the confessor's relationship with God.

The beginnings of this linguistic elision between the act of talking about an inner state and that inner state being shared can be seen in a 1903 novel by Edith Wharton, *Sanctuary*, in which she writes, 'Her smile, however, was prolonged not so much by his approach as by her sense of the impossibility of communicating her mood to him. The feeling did not disturb her. She could not imagine sharing her deepest moods with any one' (Wharton, 1903: 6–7). For the first time in the corpus, the concepts of communication and sharing are explicitly linked. Here, rather than having her deepest moods shared with another, what the protagonist cannot imagine is *telling* another about her deepest moods, and hence most personal and intimate feelings. Yet another novel from around that time also suggests that sharing confidences is the act of talking itself, and not that talking produces shared confidences. The protagonist, Jessica, writes to a priest as follows: 'I could never reconcile myself to the incongruity of confessing in our experience meetings. It seemed to me that sharing my confidence with so many people was heterodox to nature itself' (More and Harris, 1904: 115). The act of sharing knowledge is inextricably (and possibly ironically) linked to speech in a novel from 1917, when a schoolboy describes his teacher as 'extremely companionable and loquacious. He had a passion for sharing with others whatever knowledge he had, or simply hearing himself speak' (Cahan, 1917: 92). Sharing knowledge is becoming an act of communication per se, and not just the consequence of the analytically distinct practice of talking.

A slightly later text shows an even closer merging of 'sharing' and 'talking'. In this excerpt, a woman working in a factory is bemoaning how hard it is to find time for conversation at work, especially conversation about intimate matters: 'It was always difficult carrying on a conversation with Ada. She was being hollered for from every corner of the factory continually, and in the few seconds we might have had for talk I was hollered for. Especially is such jumpiness detrimental to sharing affairs of the heart' (Parker, 1922: 237). Not only does this use of 'sharing' equate sharing affairs of the heart with talking about them, it also posits a set of conditions that are best suited for such talk – and the noisy and demanding factory does not represent them.

By the 1930s, it would seem that the word 'sharing' was increasingly freely used to refer to a kind of speech. In a piece in *Harpers* magazine about the young French man, for instance, we learn that 'He acquires too the habit of talking freely with a woman, of sharing his worries and his triumphs with her' (Dunbar Bromley, 1932). Similarly, in a textbook on education, sharing experiences is represented as an act of communication: 'The primitive human group, limited to gestures, pantomime, inarticulate sounds, and crude illustrations as the only means of sharing experiences, were capable of but little progress. The achievement of an articulate language [...] made possible the oral transmission of accumulated information' (Thomas and Lang, 1937: 144). Language, in other words, makes it possible to share information. The shift to sharing as an active practice is seen in a 1940 book about community centres, where we read that in 'the active sharing of ideas the participants acquire the experience of functioning as a group' (Butler, 1940: 90). Here, then, it is not that ideas are shared in the sense of being common to the participants, but rather participants actively share their ideas with one another; that is, they communicate them. The same idea is conveyed in a letter written by Mahatma Gandhi in 1944 to the Viceroy of India, in which he writes, 'I thought that, if I claim to be a friend of the British, as I do, nothing should deter me from sharing my deepest thoughts with you' (n/a, 1944). Here 'sharing' clearly means 'telling'. Moreover, it is a friendly kind of telling; because Gandhi sees the British

Viceroy as a friend, it is most appropriate for him to share his deepest thoughts with him.

This usage is even clearer in the following extract from a 1950 novel. Here, a group of people are talking about a recently deceased friend: 'This was the atmosphere which should have been created at the luncheon after the funeral, the dear comfort of speaking of the dead, of rehearsing their virtues, of sharing one's feelings about them, Saul thought' (Sarton, 1950: 220). Saul then goes on precisely to share his feelings about the deceased; that is, he talks about her. From the 1950s, educationalists also started talking about sharing as a kind of telling. For instance: 'Time spent in "show and tell" or other types of sharing periods serves a real purpose in developing pupil interest in each other as well as opening up areas for more intensive study. Using class time for the sharing of problems when they desire it is a way of increasing pupils' willingness to help each other' (Wiles, 1952: 92). Another indication that 'sharing' had come to mean 'talking' is offered in a 1956 novel, in which one of the characters, who is worried about being overheard, 'closed her lips against sharing' (Chute, 1956: 108).

This usage – with which we are certainly familiar today – would seem to be well established by the mid-1970s. In Philip Roth's *My Life as a Man*, for example, we learn that Moe has a predilection for talking politics to his young sons: 'ever since they could understand an English sentence, he has been sharing with them his disappointment with the way this country is run' (Roth, 1974: 122), where 'sharing' obviously means talking to them about his disappointment. That same year, in an article about a support group for women with breast cancer, we read that, 'Sharing her innermost thoughts, one lovely woman spoke of being so proud on her wedding night to give her husband a full and healthy body' (Cole, 1974: n.pag.). Also in 1974, a book about education made numerous uses of the word 'sharing' in its increasingly familiar sense of 'telling', or 'communicating'. For instance, we read of 'the opening of communication and sharing between teachers and between teachers and others' (Gross and Gross, 1974: 8), and we are also presented with the idea that 'Simply sharing feelings – in a relaxed, supportive setting – can be the start of [...] authentic communications' (p. 214). In

1980, two characters in a novel are described as 'talking far into the night and sharing their experiences' (Krantz, 1980: 185). In a Dean Koontz novel from 1983, he writes that 'the other inmates were sharing news cell to cell, but none of them would share anything with Kale', who for his part 'didn't ask anyone to share information with him' (Koontz, 1983: 217).

The specifically therapeutic context of sharing also emerges at around this time. In an article in *Good Housekeeping* magazine about the hostility between doctors and nurses in a hospital, we are told how a therapist was brought in to resolve the tension. The therapist 'began the session by explaining that he hoped to help them overcome their difficulties in communicating with one another. He asked everyone to relate a particularly moving experience from childhood. The doctors and nurses took turns sharing an acute emotional moment from their lives' (Morris, 1983: n.pag.). Critical, not to say mocking, approaches also emerge. *Newsweek*'s TV critic, Harry F. Waters, wrote a piece about what he called the 'theater of therapy' – the new American talk shows in which people seemed to have no problem 'sharing such intimate details of their compulsions and dysfunctions with a total stranger' (Waters, 1983: 116). In less sarcastic register, an entirely serious article about support groups – which were booming in the early 1980s (Wuthnow, 1994) – explains that 'The therapeutic value of sharing grief with others is behind the surge in self-help groups' (Maloney, 1983: n.pag.). A gendered preference for the type of intimate communication we now associate with the word 'sharing' is discussed in a *New York Times* article from 1989, in which we learn that: '"Girls' friendships tend to be more intimate than boys'," Dr. Berndt said. "They're more concerned about sharing thoughts and feelings"' (Kutner, 1989: C8).

By the 1990s, the new meaning of 'sharing' as 'telling' had become fully embedded in everyday language. In this quote, for instance, the word is used without it being given a second thought: 'Storytelling is as natural as breathing, as old as the stone age and as current as Garrison Keillor. People are always sharing stories, whether neighbors in the car pool, business associates over lunch or children at day's end' (Elwood, 1993: n.pag.).

If the word 'sharing' has come to mean a type of 'telling', nowadays it also pertains to the conveying of information: we shall look at sharing as the constitutive activity of social media more closely in the next chapter. The *Corpus of Historical American English* allows us to trace this notion of sharing information, which, as shown in Figure 2.2, is a relatively new usage of the word. Indeed, the first instance in the corpus in which we encounter the sharing of information is from 1926 in the business context of a discussion of 'the sharing of full information' with shareholders (Ripley, 1926). Also in a business context, a guide book for the office supervisor extols the value of 'Showing an interest in other people's work and seeking information and sharing it' (Niles and Niles, 1935).

During the 1940s and 1950s, the primary context in which 'sharing' and 'information' went together was that of science, and especially nuclear science. In other words, information sharing was a practice found at the institutional level, and what was shared were business or scientific data. In the 1960s, individuals were described as sharing information, though in the excerpts from *COHA* this is not very personal information. Indeed, although information sharing was seen as a practice that 'encourages trust among employees' (Cole, 1983: 22), it was not until 1995 that the sharing of information was seen as in any way intimate: 'Our sense of self and friendship are determined by control of personal information', we read in an article about online privacy. 'The sharing of information is intimate', concludes the journalist (Harrington, 1995: F14). Mostly, though, in the extracts from *COHA* that relate to information sharing, the information is very often technical, and those doing the sharing tend to be institutions. This kind of sharing appears to lack the normative aspect of sharing-as-telling as just described.

The concept of knowledge sharing shows a similar pattern. It was not a turn of phrase that one would hear or read in the nineteenth century; indeed, the first such usage in the corpus is from 1917, in an extract cited above ('He had a passion for sharing with others whatever knowledge he had' (Cahan, 1917)). Later instances portray sharing knowledge in a positive light. For instance, an artist is quoted as saying that 'Sharing knowledge with others is important and enriching' (n/a, 1999); or: 'Sharing knowledge with others reinforces the knowledge within us' (Haddon, 2000: n.pag.).

Another important stream, in which the two previous streams can be seen as merging, is the growing association of 'sharing' with intimate relations and a vague sense of goodness and fairness. The progression of the word throughout the twentieth century shows talk and intimate relations to be increasingly interrelated, as 'sharing' came to be defined as a central pillar of close interpersonal relations. An early indication of this comes from a description in 1846 of a mother, who is said to be an 'affectionate and faithful parent' on account of her 'sharing with [her children] whatever she might have' (n/a, 1846: 462). The idea of commonality is clearly central to these ties, and is also expressed in an 1865 article in *Harpers* about marriage in China: 'Eating from the same sugar cock and drinking wine from the same goblets, are symbolical of union in sharing their lot in life' (Doolittle, 1865: 435).

An excellent example of 'sharing' being held up as a pillar of marriage is found in an article published by anthropologist Margaret Mead in *The Nation* in 1953. There, she wrote that, 'Since World War II a new kind of marriage has developed in America', and then went on to list some of its attributes: it has 'greater frankness, greater articulateness, greater sharing than any we have known before in this country' (Mead, 1953: n.pag.). It is interesting to note here the absence of an object of sharing. Mead does not say *what* is shared (confidences/interests/the housework/money?), but just that the modern marriage is characterized, *inter alia*, by 'sharing'. The description in a novel from roughly the same time of 'the fun of marriage' as lying in the 'sharing of things with your mate' is hardly less vague (McCarthy, 1955: n.pag.), but this is exactly the point: during the second half of the twentieth century 'sharing' comes to stand for the qualities of the ideal marriage or, indeed, for the 'pure relationship' (see Giddens, 1992). By 1973, in a sociological analysis of the family, the author takes for granted 'the modern values of sharing and reciprocity in marriage' (Yorburg, 1973: n.pag.). By the 1980s, we read texts such as this: 'I saw the sharing that they had together as very important in marriage' (Marks, 1986: 22), or this: 'Marriage means together, it means sharing and loving' (Leib, 1989: 284). This idea continues through to the present day. In a book on marriage from 1995, for instance, the authors report receiving a letter with the following statement: 'This is what marriage is to me, the sharing of two lives to complete each other' (Wallerstein and Blakeslee, 1995: 4). This cultural conception of marriage as based on sharing also occurs – though anachronistically so – in a piece of fiction written in the twenty-first century that is set in 1899: 'That's what a marriage is – sharing. You share a home and a place. You share children' (McGraw, 2006: 528).

What is notable about these last examples is that while the word 'sharing' is a metaphor, its meaning stands on its own and it no longer brings to mind its non-metaphorical senses at all. Linguists might say that in this instance the metaphor is actually 'dead' (on the death of metaphors, see Billig and MacMillan, 2005). The criteria for deciding whether or not a metaphor is dead are to do with the relationship between the metaphor and its literal counterpart (Deignan, 2005). For instance, when used in the phrase, 'I saw the sharing they had together', the word 'sharing' does not evoke its literal sense. In Deignan's terms, in these uses of 'sharing' to denote what is good about relationships its sense is not dependent on the literal meaning; the literal sense is not 'core' to the metaphor (Deignan, 2005: 41–7). We know what kind of relationship is being referred to if it is described as one of 'sharing': there is good communication between the couple (Katriel and Philipsen, 1981); they are considerate towards one another; and – most importantly for the new sense of sharing that emerged during the 1980s – their relationship is based on equality and fairness, trust and mutuality. This new sense, then, in turn becomes the source of a further metaphorical development when, as we shall see, the concept of sharing in social media is presented as good for interpersonal relationships.

Sharing and Caring

One particular combination of words that captures the sense of sharing that I am driving at here is that of 'sharing' with 'caring'. The Google Books Ngram Viewer shows a leap in the prevalence of the phrases 'sharing and caring' and 'caring and sharing' in the 1960s, which flattened out only around the year 2000 (see Figure 2.3). The conjunction of 'sharing' with 'caring' thus clearly overlaps with the emergence of the hippie movement and the American counterculture more generally – a point that I will come back to in the following pages.

Clearly the rhyme plays a part in the popularity of the phrase (Jakobson (1960) analyses the popularity of the slogan 'I like Ike'; McGlone and Tofighbakhsh (2000) present the findings of experiments that show people are more likely to agree with the message of an aphorism when it rhymes). However, while the rhyme was always present, the phrase was not. Indeed, of the thirteen instances from *COHA* in which the words 'sharing' and 'caring' are adjacent or have only one word between them, all but one are from 1982 onwards.[6]

The modern sense of the term 'caring and sharing' (or 'sharing and caring') enacts a warm glow around sharing. We can see this sense, for instance, on the cover of a Care Bears book, where the phrase 'Caring and Sharing' is accompanied by teddy bears, fluffy clouds and flowers with rainbow-coloured petals (Figure 2.4). For good measure, the purple bear is none other than Share Bear himself, and one of the bears – Cheer Bear – has a rainbow on her stomach. A similar visual representation of sharing is found on Dropbox's website. There, the image that is chosen to appear alongside the word 'Share' is that of a rainbow (Figure 2.5). The rainbow is of course a powerful symbol that represents harmony and peace. In many cultures the rainbow is seen as bridging our world and the heavens. In Judaeo-Christian cultures the rainbow's symbolic force comes from Genesis 9: having flooded the earth, God promises never to do such a thing again, and the rainbow is the symbol of this covenant. The rainbow also has connotations of spirituality, which resonate with the idea of sharing as a counterbalance to thoroughgoing materialism. In any case, this kind of visual trope is familiar to us from schmaltzy greeting cards and from the tone of voice we are likely to adopt when we say the very phrase, 'sharing and caring'.[7]

Returning to written texts, in 1983, in response to the emergence of the so-called 'New Man', a book maintains that

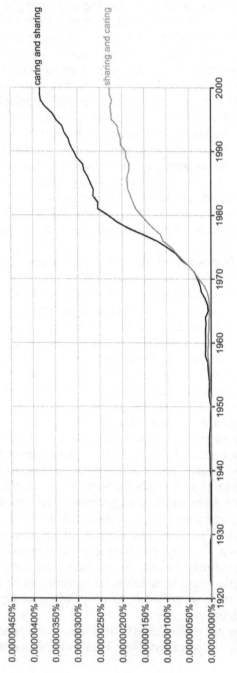

Figure 2.3 Sharing and caring, 1920–2000

Figure 2.4   Care Bears; © Scholastic

 Files

 Photos

 Sharing

 Links

 Events

Figure 2.5   Dropbox icons; Dropbox

'men are being given permission to be caring, sharing, domestic, and paternal' (Kron, 1983: 121). Michael Ignatieff adopts a somewhat ironic stance towards the phrase when he writes that 'the moral virtues of "sharing and caring" are increasingly valued in American culture because they are healthy, because the immunologists say that carers and sharers' coronary statistics are impressive' (Ignatieff, 1988: 28). Not dissimilarly, a retail consultant talking about tween girls and brands they prefer is quoted in an interview as saying that girls 'have a caring, sharing and compassionate attitude' (O'Donnell, 2007: 1B).

As a final example, which I propose to analyse more closely, let us turn to Tiger Woods' book *How I Play Golf* (2001). In the introduction, Woods writes that the book has been written 'as the ultimate tribute to Mom and Pop's ideal of caring and sharing'. Here, the phrase simply means that his parents were good, considerate, unselfish people. Fortunately, though, Woods goes on to detail his understanding of what 'caring and sharing' means: 'In essence, if you care for someone you'll share with them your most treasured possessions.'

As I think is clear from the previous instances of 'caring and sharing', the idea of sharing treasured possessions does not really appear to belong here. Sharing – in the context of caring and sharing – usually refers to intimate communication between people, or to the fact of their sharing experiences, or even their lives, however we might understand that. In fact, we could go as far as to say that none of the other instances of 'sharing', where that word is used to describe a quality of an intimate relationship, is about possessions. And, indeed, in the very next sentence, Woods tells us what it is that he is sharing with the reader and, while it is something he may very well treasure, it is hardly a possession. He writes: 'In this book I will share with you a lifetime, albeit a relatively short one, of knowledge about the greatest game in the world' (Woods, 2001: 15). In this instance, the word 'sharing' is doing a lot of the work that this book seeks to analyse. It implies a positive model of interpersonal relations; it implies an altruistic foregoing of at least part of a 'treasured possession' in a zero-sum game; and it implies the conveyance of information in a non-zero-sum game. One can almost see the different senses of 'sharing' sparking one another off in Woods' mind as he writes (or in the mind of his ghost writers – but the point still stands): his mom and pop believe in 'caring and sharing' as the somewhat vague ideal on which they base their interactions with others, and especially their parenting practices; 'sharing' and 'caring' are linked because if you care for someone, then you share with them, but for this sharing to have affective meaning, it must entail some kind of sacrifice, hence the reference to 'treasured possessions' (though they are only being shared, not given away); it then turns out that what Woods wants to share is knowledge, but this sharing is now imbued with affect through the word's previous association with 'caring' and Woods' mom and pop. Were we of a cynical bent, we might say that all this talk of sharing actually misconstrues what is going on here. Woods may well be sharing his knowledge with us – this, as we have seen, is a well-established usage of the word 'sharing' – but it is hard to see what this has got to do with 'caring'. First, Tiger Woods does not know the people reading his book, which makes his talk about 'caring' ring somewhat hollow. And, second, while Woods is sharing his knowledge, he is also making money. The idea that his relationship with his readers is equivalent to his parents' relationship with him, in that both are based on 'sharing and caring', is pretty far-fetched.

Having said all that, my point is not that Tiger Woods is being manipulative, nor indeed that Facebook is being manipulative when it says it enables us to share with the people we care about. My point, rather, is that these are exactly the associative chains that are formed in our minds when we talk about, and hear others talk about, sharing. These are the associations that are culturally available to us today. In saying this now I am pre-empting an argument I shall make later about the term 'file sharing'. One of the arguments against the term 'file sharing' is that it is 'not really sharing' because the people involved are not physically co-present, listening to music or watching a TV show together. While I think that 'file sharing' is an adequate term that captures its subversive nature vis-à-vis the current copyright regime, my more general stance is not that of language policeman, stipulating what may or may not properly be called 'sharing'. For a start, while we may see Tiger Woods as making money from sharing his knowledge about golf with us (and some people may prefer to put 'sharing' in quote marks to express their cynicism (Predelli, 2003)), he himself may feel a sense of community with his readers as he imparts knowledge that is clearly of personal significance to him. In other words, he may feel as though he *is* 'sharing', in its 'caring and sharing' sense. One interesting speculation, which is probably impossible to prove, but which is hinted at above, is that Tiger Woods feels as though he is sharing precisely because of the word's semantic associations, as represented by the phrase 'sharing and caring'. To reiterate: it is these associations between semantic fields that make it possible for Dropbox to place a rainbow next to the word 'Share' on their website, and that provide the inspiration for the cartoon in which buying beer for a child is wrong, but sharing it with him is fine (once more with rainbow-like colours; see Figure 1.1).

Conclusion

In this chapter, I have shown how 'sharing' came to be a type of talk in addition to its older meaning as a form of distribution. We shall come back in Chapter 5 to the centrality of sharing to therapeutic culture; here, let us note that sharing became telling precisely around the sharing of emotions. In the nineteenth century, one might have shared one's troubles, and one would have done so by talking about them. It was only in the first half of the twentieth century, though, that the act of talking about one's troubles (or grief, or sorrow) was elided with the outcome of that talk (one's troubles were shared, that is, distributed among other people, thereby making them lighter, according to the distributive metaphor). At the same time, talking in general was becoming more important and increasingly constitutive of the self and its relations with others (Illouz, 2008). Sharing, as a particular type of talk (Carbaugh, 1989), became integral to intimate relationships.

The schmaltzy, rainbow-coloured qualities of sharing would appear to be a strong statement of what sharing is not, or of the opposite of sharing. If not-sharing is selfish, unsympathetic and inconsiderate – which for some may describe the harsh world of capitalism – then Care Bears and puppies and two children eating from the same ice-cream cone would appear to represent all that is good about the world. It was something like this that John Lennon meant in his song 'Imagine', when he sang about all the people sharing all the world. This is not to say that some of us have not developed a cynicism towards this – I myself have just called it 'schmaltzy' – but it is to establish what associations we have when we think about sharing. In the next chapter, we shall take these insights into the layers of meaning of the word 'sharing' and critically analyse the centrality of sharing to social media.

3
Sharing and the Internet

Manuel Castells has posited that, '*In our society, the proto-cols of communication are not based on the sharing of culture but on the culture of sharing*' (Castells, 2009: 126; emphasis in original). Similarly, Sherry Turkle has said, in critical mode, that the 'best way to describe' the 'new regime' brought about by contemporary communication technologies, and especially smartphones, was 'I share therefore I am.'[1] These are but two examples of the way in which the notion of 'sharing' is deployed quite naturally nowadays in the context of digital communication.

For some, the idea that what we do online is sharing is not new. For instance, in a 2006 blog post introducing the 'Share on Facebook' bookmarklet, Chris Hughes, Facebook's 'manager for Share', wrote: 'Ever since this whole Internet thing got started, people have been sharing stuff left and right.'[2] Similarly, though writing from an entirely different political position, self-described 'venture communist' Dmytri Kleiner has stated that 'The internet has always been about sharing between users' (Kleiner, 2011: 179).

According to popular wisdom and rhetorical convention, then, the internet is and has always been about sharing.[3] However, inspection of some of the literature about the emergence of the internet, and an examination of the way that SNSs presented themselves during the years of their emergence and consolidation (approximately 2000–10), raise three important observations. First, the cultural and discursive association between the internet and the values of sharing is not a natural given. Tarleton Gillespie's observations concerning the term 'platform' are equally apt here: 'A term like "platform" does not drop from the sky, or emerge in some organic, unfettered way from public discussion' (Gillespie, 2010: 359). This, of course, raises the question of where it does come from. Second, nor is the internet-sharing linkage a direct consequence of the architecture of the internet; instead, as Thomas Streeter argues, what matters is what 'we have brought to the internet rather than what the internet has brought to us' (Streeter, 2011: 187). Third, people have *not* been 'sharing stuff left and right' ever since 'this whole internet thing got started'. They may have been posting stuff, uploading stuff and sending stuff, but those actions were not called 'sharing'. Later in the chapter, I shall show when they became 'sharing'. As I have already said, if we live by our metaphors, then this is significant and deserves attention.

This chapter gets under way with the acknowledgement that, since its inception, the internet has been culturally and discursively associated with collaboration, cooperation, connectivity and community.[4] This is reflected in a strong utopian tradition of writing about the internet, which includes authors such as Howard Rheingold (1993), Clay Shirky (2008), Charles Leadbeater (2008) and *Wikinomics* authors Don Tapscott and Anthony Williams (2006). It also includes such varied texts as Yochai Benkler's *The Wealth of Networks* (2006), Nicholas Negroponte's *Being Digital* (1995) and Bill Gates' *The Road Ahead* (Gates, Myhrvold and Rinearson, 1995). The reader will doubtless be able to think of plenty more examples. I then move on to the observation that even if the internet has always been associated with prosociality, this prosociality has not always been subsumed under the concept of sharing – except in one important instance, namely, hacking culture. In order to demonstrate this I shall spend some time talking about people not talking about sharing – a tactic to which I return in my discussion of file sharing in Chapter 6. I then pay close attention to the emergence of sharing as the keyword for social media and what was, for a while, called Web 2.0.[5] At this point – as the new metaphor is taking root – we can see people engaging with it, testing it, playing with it. We can also see people extrapolating it backwards and applying it to a time when people did not actually talk about the internet in terms of sharing. Taken together, these issues lead us once again to ask about the special power of 'sharing' as the main internet metaphor today.

Constructing the Internet as Prosocial

The meanings of new technologies are not presented to us ready for consumption; they do not come bundled with the technology itself. Instead, meanings are attributed to technologies by specific groups that occupy specific social positions. Science and technology studies (STS) researchers call this 'interpretive flexibility' (Pinch and Bijker, 1987). Examples are manifold: in a widely cited article, Kline and Pinch (1996) mention Susan Douglas on the radio (1987), Claude Fischer (1992) and Michele Martin on the telephone (1991), and David Nye on electricity (1990). Even histories of the internet written very close to its explosion into the public consciousness emphasized quite different aspects, as elegantly detailed by historian Roy Rosenzweig (1998), who compares a number of very divergent accounts of the creation of the internet (such as the Haubens' *Netizens* (1997) and Paul Edwards' *The Closed World* (1996)).

While not the only book to illustrate the assignation of meaning to the emergent internet,[6] Fred Turner's *From Counterculture to Cyberculture* (2006) demonstrates a clear instance of this process. Specifically, he sets himself the task of explaining how 'the cultural meaning of information technology [shifted] so drastically' from the 1960s, when computers were seen as an oppressive tool in the hands of an oppressive society, to the 1990s, when they were an integral part of a 'countercultural dream' spearheaded by a disparate group that Turner calls 'the New Communalists' (p. 2). 'How was it', he asks, 'that computers and computer networks became linked to visions of peer-to-peer adhocracy, a leveled marketplace, and a more authentic self?' (p. 3). Note that these visions approximate present-day conceptualizations of sharing. Note also that Turner does *not* ask how computer networks became linked to visions of sharing.

Significantly for the later adoption of the concept of sharing, and as will be shown in Chapter 5 on the therapeutic discourse, the counterculture described so lucidly by Turner was based on a type of interpersonal relationship with others as well as representing a state of self. For instance, the features of the internet that Turner seeks to explain include its purported ability to 'render the individual psychologically whole, or drive the establishment of intimate [...] communities' (p. 3). This is seen especially clearly in Turner's treatment of the Whole Earth 'Lectronic Link, or the WELL, an online forum which was the first to be described as a 'virtual community' (Turner, 2006: Chapter 5). The WELL was set up as a non-hierarchical, self-governing system and was populated, among others, by former dwellers of the communes of the 1960s and 1970s. Howard Rheingold was an early and influential member, and his classic text, *The Virtual Community* (Rheingold, 1993), was based on his experiences there. For him, computer networks offered the ability to 'rediscover the power of cooperation' and to 'recapture the sense of cooperative spirit that so many people seemed to lose when we gained all this technology' (Rheingold, 1993: 109). This is a postlapsarian take on community (which we shall encounter again when we discuss the sharing economy in the next chapter) and has a familiar structure: at some unspecified time in the past, we used to know what the power of cooperation was, and we used to have a cooperative spirit, but 'all this technology' robbed us of that knowledge – 'While we've been gaining new technologies, we've been losing our sense of community', states Rheingold (p. 109); now, though (he said in 1993), people are learning to 'use computers to cooperate in new ways' (p. 109).

The force of Turner's work is in helping to understand how someone like Rheingold came to his views regarding the place of computer networks in human sociality. As such, their texts are quite differently related to the activities on the ground: Rheingold's is a first-hand account; Turner's is an effort to contextualize Rheingold's account (and many others like it). By doing so, Turner makes an important contribution to our understanding of the cultural processes behind one of the major imaginaries of the internet today (see also Mansell, 2012), an imaginary that enables us to understand what is meant when people talk about the internet as having a culture of sharing.

However, something that both Turner and Rheingold have in common is talk that strongly resonates with the values of sharing but that actually makes very little use of the term 'sharing'. The metaphor was not as readily available to them then as it is to us now. In other words, the internet was framed (by Rheingold and his peers) as a technology for collaboration and cooperation, which lent itself very readily to the metaphor of sharing. However, talk of the internet as a platform for 'sharing' is not present in Rheingold's account of the WELL, nor in Turner's account of the internet's associations with the American counterculture. As I shall show below, 'sharing' became the keyword for the internet in the mid-2000s, concurrently with, and probably as a result of, the widespread adoption of the term by the hugely expanding SNS scene. Even so, in the 2000 edition of *The Virtual Community*, though not in the original 1993 version, Rheingold paraphrases an article by Barry Wellman and Milena Gulia (1999) as arguing that 'cyberspace is a place where sharing is encouraged' (Rheingold, 2000: 364). Interestingly, though, Wellman and Gulia actually make no mention of sharing at all in that paper.[7] We might see this as a small moment in the process of the internet becoming a place of sharing. A similar moment can be seen in Turner's discussion of the WELL, which he describes as having 'an emphasis on sharing, intimacy, and leveled social hierarchies' (p. 248), while in Rheingold's account of the WELL there is actually precious little talk of sharing at all.

Moving from the internet in general to social network sites in particular, I offer one final example of the emergence of 'sharing': in their canonic article, boyd and Ellison (2007) define and analyse SNSs with very few mentions of sharing (there are some references to media, information, video and picture sharing – what I shall define below as sharing with concrete objects). However, fast-forward to their 2013 contribution to the *Oxford Handbook of Internet Studies*, which in many ways is an update to their 2007 article, and now there is a great deal of talk of sharing: people 'create content to share with their contacts' (Ellison and boyd, 2013: 154); we are told that the 'primary driver of SNS use' is 'The desire to communicate and share content' (p. 159); we also learn that SNSs have lowered 'the barriers to communication and sharing' (p. 159). The clearest example of the point I am trying to make, though, is when Ellison and boyd are laying out their 'Definition 2.0' of social network sites. 'Social network sites have evolved,' they say, 'but their foundational activities – sharing content with a bounded group of users – are fundamentally the same' (p. 159). In the 2007 paper, though, not only does the term 'sharing content' not appear, but the emphasis is far more on profiles than on sharing (as Ellison and boyd themselves acknowledge in 2013). To be clear, my purpose here is not to quibble with boyd and Ellison's crucial contributions; rather, it is to use these two important texts to show that we have not always talked about SNSs in terms of sharing and, moreover, that there is a tendency to retroactively affix the term 'sharing' to practices that were not necessarily called that at the time.

This is not to say that talk of 'sharing' was not associated with early computing use: as I shall show in Chapter 6 on file sharing, the early computer systems worked on technologies that were called 'time sharing', and subsequently 'disk sharing'. This use of 'sharing', though, was quite literal: the mainframe's time was being divided up between the machine's users; the disk was being shared by users in the same way that housemates might have a shared dishwasher. In other words, the terms 'time sharing' and 'disk sharing' had no normative connotations. In relation to hackers, however, things are different. Here, the concept of 'sharing' would appear to have been central right from the start.

Sharing and the Hacker Ethic

In Steve Levy's classic account of hackers (Levy, 1984), right from the start he defines the 'hacker ethic' as 'a philosophy of sharing, openness, decentralization, and getting your hands on machines at any cost to improve the machines and to improve the world' (p. ix). The value of sharing seems to have been part of the community's self-conception as well: in the very initial discussions around the formation of the legendary Homebrew Computer Club in March 1975 (the site of the first notorious case of software piracy), the words that came up most were 'cooperation' and 'sharing', reports Levy (p. 202), and the group would develop a 'time-honored practice of sharing all techniques, of refusing to recognize secrets, and of keeping information going in an unencumbered flow' (p. 276). Leader of the club, Fred Moore, placed sharing central to the experience of being a hacker: 'By sharing our experience and exchanging tips we advance the state-of-the-art', he wrote in a newsletter (p. 214). As another early Homebrew member told Levy later on: 'More than any other individual, Fred Moore knew what sharing was all about [...] That was one of the expressions he was always using – sharing, sharing, sharing' (p. 214).

There is other textual evidence that points to the centrality of sharing to hackers and the hacker ethic. In *The New Hacker's Dictionary* (Raymond, 1996), for instance, the 'hacker ethic' is defined as: 'The belief that information-sharing is a powerful positive good, and that it is an ethical duty of hackers to share their expertise by writing free software and facilitating access to information and to computer resources wherever possible' (p. 234). This definition itself contains at least two senses of sharing: sharing information implies telling others about information you have, while sharing expertise means freely distributing the fruits of your hacking labour and helping others to access resources. Sharing is also central to the hacker ethic as described by Pikka Himanen (2001), who adds to Raymond's definition that hacking 'should be motivated primarily not by money but rather by a desire to create something that one's peer community would find valuable' (p. x). Referring to the hacker ethic as a work ethic in the same sense that Weber described the Protestant work ethic opens a pathway to what is now commonly known as the sharing economy – this is a path pursued later in this book.

The two other contributors to Levy's book also write about sharing. Linus Torvalds (creator of Linux) wrote that 'The reason that Linux hackers do something is that they find it to be very interesting, and they like to share this interesting thing with others' (p. xvii). And in the book's epilogue, Manuel Castells discusses the emergence of a 'new culture' characterized by the 'augmentation of innovation potential by cooperation and sharing' (p. 177), using the term here in a general sense that stands in contradistinction to capitalism.

In the field of hacking, then, sharing is and has been a central concept, where it is used to stand for anti-capitalist thought and practice: hackers share the products of their labour, they do not sell them; they share their expertise rather than charge an hourly rate for it; they share their knowledge and interests for the sake of forging community and making the world a better place. This is different from more general appraisals of the internet as a whole, in which the internet was talked about in terms of the values of sharing but without the word itself. I account for this in terms of the *digital* nature of hacking, as suggested above. The original hackers, who were deeply embedded in the computer scene, were already familiar with time sharing, disk sharing and file sharing, terms which, as noted, had little or no normative connotations. It would not have taken much linguistic imagination, though, to add the positive valence of sharing from other contexts to the idea of information sharing. We shall look at this more closely in the discussion of file sharing in Chapter 6.

If Not Sharing, Then What?

If the metaphor of sharing was not employed by commentators – both academic and lay – of the internet, how *was* its prosociality conceptualized? Before we were 'sharing' online, what was the label given to what we were doing?

One answer to this question is 'gifting'. For a while, acts of online prosociality, generosity and otherwise unpaid-for activities (such as creating a homepage, writing a blog or opening your music library to other Napster users) were analysed in terms of gifting, with Marcel Mauss's seminal work (1966 [1925]) providing inspiration for a tranche of articles that tried to understand why people would do stuff online for free.[8] This can be seen in Turner's book (2006: esp. p. 157), in which he adopts Rheingold's conceptualization of the WELL as 'a kind of gift economy' (Rheingold, 2000: 49). This understanding has been evidenced in a number of articles in the online journal *First Monday*, with Richard Barbrook's influential article on 'The Hi-tech Gift Economy' (Barbrook, 1998) the stand-out piece (but see also Bays and Mowbray, 1999; Ghosh, 2005; Stalder, 1999; Veale, 2003). Another notable example is Markus Giesler's (2006) analysis of Napster as a 'consumer gift system'. In this article in particular what is somewhat surprising, and yet easily understandable, is the insistence on the conceptualization of gifting when the concept of sharing would seem to be just waiting to be picked up. This is not a criticism of Giesler's scholarship – my point is merely to highlight the relative novelty of sharing – but let us notice how Giesler talks about sharing: for instance, he quotes an interviewee who explains that 'everything that is shared is just accessible by everyone else' (Giesler, 2006: 286); and later on he explicitly talks about the 'social discourses, practices, and structures of sharing' (p. 287) even as he furthers his analysis of file sharing as a gift economy. Because my interest here is in the rise of a metaphor, I shall not engage with the question of whether Napster was an instance of a gift or a sharing economy, and in any case I shall discuss sharing economies and file sharing at length in subsequent chapters. What I have suggested here is that before the internet became all about sharing, its prosociality was conceptualized in terms of gifting and the gift economy.

Today, of course, discussions about the internet are suffused with 'sharing' talk, which brings it within the larger metaphorical structure of sharing that this book seeks to unpack. This explicit association between the internet and sharing is obviously in need of an explanation. My argument is that it is to do with the rise of SNSs and the central place the rhetoric of sharing has in them. It is to the rise of sharing in social media that we therefore turn.

Sharing and Social Media

Sharing is the constitutive activity of social media.[9] It is the umbrella name given to the myriad activities we carry out online: updating statuses; uploading photos and videos; writing reviews on books and other products on Amazon and elsewhere; tweeting; checking in; and in fact almost anything we do. But why 'sharing'? What is the rhetorical force of the word in this context? How is it used? Have its uses changed over time? What is the political economy of 'sharing'? And does the word itself serve any ideological purposes?

In this section I focus on the evolution of the word 'share' in the context of SNSs with the objective of historicizing it and charting its emergence as the descriptor for our online participatory activities. This is an important task given the centrality of SNSs in our lives today, which in turn is conveyed in reflections on the word 'share' in popular media outlets and, as we shall see later on, in its appearance in the phenomenon of the 'sharing economy'. If one of the objectives of this book is to unravel the movement of the metaphor of sharing through different spheres, we should clearly take some time to unpack its meanings in the context of social media.

The richness of the word in this regard has not escaped the attention of columnists and critics. English satirist Charlie Brooker, for instance, responded to the launch of 'frictionless sharing' between Spotify and Facebook by describing sharing as 'a basic social concept that has somehow got all out of whack'.[10] 'The idea behind sharing is simple', he writes. 'Let's say I'm a caveman. I hunt and slaughter a bison, but I can't eat it all myself, so I share the carcass with others, many of whom really appreciate it.' Here Brooker draws on the trope of the hunter-gatherer sharing his prey – a theme explored in more detail in Chapter 4 – where the logic of this kind of sharing is a distributive one. 'But it's not all bison meat', Brooker goes on. 'The other thing I share is information.' Here, sharing has a communicative logic. According to Brooker's satirical take on current-day sharing, sharing among cavemen was functional: 'It kept the community fed, as well as entertained and informed.' These observations then serve as the springboard for a critique of today's economic inequalities (the rich do not share) and trends towards greater sharing: 'Not sharing money or bison meat, but personal information.' As I shall argue shortly, the ease with which Brooker can shift between different meanings of sharing reflects a feature of the word that has made it extremely useful for SNSs, namely, its diversity of uses and logics or, put differently, its polysemic homonymity. Brooker also puts on display some of the associations readily made with 'sharing': cavemen/hunter-gatherer societies; the greed of bankers as the opposite of sharing (also referenced in the opening vignette of Benkler's *Sharing Nicely* (Benkler, 2004)); and telling people what you are thinking and doing. Notably, it is these associations that form the basis of Brooker's critique, according to which Spotify updating Facebook as to what its users are listening to should not properly be considered sharing; at the very best, it is a sign that the concept has 'got all out of whack'.

My approach here is somewhat more agnostic than that of Charlie Brooker and others who contrast SNS-based sharing with other, more 'real' types of sharing. Nonetheless, the associations made by Brooker (and many, many others) are important, as they point to the kinds of associations people make when thinking about 'sharing', including, one can assume, the people who write tag lines for SNSs. Before describing the evolution of 'sharing' (the word) in SNSs, we should note that even if people want to reproach SNSs for promoting an activity that is not 'really' sharing and wish to call foul on the appropriation of the word, it is not as if the implications of online 'sharing' are so very far away from what might be perceived as the 'truer' or more 'authentic' senses of sharing: in all cases, the act of sharing involves crossing the boundary between the private and public; it involves making accessible to a wider public material (photos, candies) that had previously been accessible only to me; and it (often) involves communicating an inner state, or status, to others. With that said, let us turn to SNSs and the brief history of the word 'sharing' in that particular context.

The website most closely identified with 'sharing' is, of course, Facebook. Indeed, Facebook itself declares that 'The power to share is the cornerstone of Facebook'.[11] Hence, it is not by chance that José van Dijck (2013) devotes an entire chapter of her book on connectivity to 'Facebook and the Imperative of Sharing'. In that chapter, van Dijck refers to two types of sharing: sharing as connectedness, and sharing as connectivity. Sharing as connectedness refers to the ways that interfaces get us to share information with others (posting statuses, uploading photos, etc.), while sharing as connectivity refers to the interactions between Facebook and third parties (through the Like button, the erstwhile Beacon service and so on). As van Dijck observes, Facebook seeks to promote the former while concealing the latter. As we shall see later on, the fact that the latter (sharing as connectivity) is also called 'sharing' (by internet companies, at the very least) has given rise to criticisms regarding the proper use of the term.

While the centrality of Facebook in what van Dijck calls the social media ecosystem can hardly be questioned, I would add that the word 'share' has a history on Facebook – as it does throughout SNSs. In fact, Facebook was not even a pioneer in its use of the word, which does not appear in any of the site's self-descriptions until 2006, by which time it was already fairly common currency on other SNSs.

Early screenshots of the Facebook homepage (when it was still thefacebook.com), for instance, enable us to re-create something of the site's self-presentation. Indeed, what screenshots from 2004 and 2005 show is an absence of 'sharing' on Facebook; instead, the site quite unimaginatively presents itself as an 'online directory' where users are told they can 'Look up people at your school', 'See how people know each other' and 'Find people in your classes and groups'. At this early stage, then, the site is for making contact with other people, but not for 'sharing' with them.

Based on a reading of Facebook's own blog posts, the term 'sharing' gained widespread use within the company during the second half of 2006, which is also the time that Facebook opened itself up to all internet users (and not only university or high school students), perhaps suggesting a relationship between the conquest by SNSs of the social and cultural mainstream and the ascension of 'sharing' as their constitutive activity. The first mention of sharing in the context of Facebook came in May 2006 in a press release to announce the site's expansion to include work networks.[12] In that press release, Facebook was described as 'the social directory that enables people to share information'. But even though the company had started to represent itself in terms of sharing, the word had not yet become part of Facebook's lexicon. This is evidenced by a blog post published under Zuckerberg's name on 30 August 2006, devoted to the changes that the site was going through (especially regarding photos, events, groups and the wall). In that post he does not mention sharing at all.[13]

The comprehensive adoption of 'sharing' by the site would appear to have been decided upon subsequently to that blog post, but before the end of October of that year, when Facebook published a blog post with the title 'Sharing is Daring'.[14] In that post, the internet is described as having always been a platform for sharing – 'Ever since this whole Internet thing got started, people have been sharing stuff left and right' – before Chris Hughes, Facebook's 'manager for Share', goes on to explain how easy it is now to share through Facebook by using the new 'Share on Facebook' bookmarklet – a button to add to one's browser that enables users to publish web content on their wall. Five days later, on 31 October 2006, Facebook announced that websites would be able to place 'links to share on Facebook'.[15] In the press release, Zuckerberg said: 'People share interesting content on the Web and on Facebook all the time. Now we're making the sharing process more efficient by giving people a simple structure to do it in.'[16] Even so, the use of the term 'sharing' was uneven across the Facebook site and was not included in the firm's tag line until 2008 ('Facebook helps you connect and share with the people in your life'[17]). Today, Facebook's mission is 'to give people the power to share and make the world more open and connected'.[18]

This is clearly marketing talk: Facebook makes money through a model of advertising based on extremely granular knowledge about its users such that the more we share, the better for Facebook. In any case, I will have more to say about the function of the word 'share' in the context of Facebook later on. For now, let us note that the word was not always integral to the Facebook experience – as that experience is marketed by Facebook, at least – and that it emerged at some point in the second half of 2006. While Facebook is currently by far the dominant player in the online social network space, it is not the only one, and it was certainly not the first. In what follows, I explore the rise and rise of 'sharing' in SNSs throughout the first decade of the twenty-first century, pointing to its increased adoption by SNSs as well as to new ways in which the word has been used. This exploration will also enable us to start pointing to one of the contradictions of the word 'sharing' in the context of social media.

What follows is based on analyses of the forty-four largest, most visited and historically significant SNSs.[19] The list of sites was compiled from three sources: the ratings of Alexa, a leading company for web metrics, for visits to websites in its social network category for July 2011; data compiled in Wikipedia on the size of the membership of the largest SNSs;[20] and the section on the history of SNSs in boyd and Ellison's survey of the field (2007; see especially Figure 1 on p. 212). Data were collected using the Internet Archive's Wayback Machine. The Wayback Machine 'is a service that allows people to visit archived versions of Web sites'[21] and has been crawling the internet since 1996.

For each SNS, I looked at the oldest available impression of their front page. I then moved forward in time, looking at each site on the first day of every month, or the date closest to that if there was no snapshot for that exact day, through to the end of 2010. I created screenshots of my own, and excerpted relevant parts of the websites. These screenshots and excerpts were imported into the qualitative data analysis software, Atlas.ti, for coding and classification.

I did not record data from every single impression I called up from the Wayback Machine, but only when a site's front page had changed from the previous month's version. Thus, changing self-representations over time can be followed, and information, such as the first time that a site presented itself in terms of sharing, could be gathered. When sites undertook a major renovation of their front page, I also visited their About or FAQ pages, assuming that the site had gone through a strategic process that might be reflected in other parts of the site as well, and given that a website's About or FAQ pages often include a longer and more detailed self-presentation than is possible on its front page. However, links off the front page did not always work, and so About, FAQ and Tour pages could not be as systematically collected as front pages. Therefore, the main arguments presented below rest on data collected from the front pages of SNSs, with other pages from those sites occasionally drawn upon to provide further examples.

Looking at the homepages of forty-four SNSs for a period of a decade reveals changing uses of the word 'sharing' alongside its notably increased prevalence. In the most general terms possible, 'sharing' in this context has come to mean participating in social media. In what follows I characterize the features of this new type of sharing and uncover their logics. In doing so, I point to three main characteristics of sharing in SNSs: the appearance of fuzzy objects of sharing; use of the word 'share' with no object at all; and deploying the notion of sharing where it was not used before.

Fuzzy objects of sharing

If I call an object of sharing *concrete*, I mean that we immediately know what is being shared. The clearest example of this is photos: when Flickr presents itself as 'The best way to store, search, sort, and share your photos',[22] we know precisely what is being shared. Likewise, when the SNS Multiply tells us to 'Share interesting web sites' (10 December 2004)[23] we understand right away that we are being encouraged to give our friends links to internet sites, and the same is true of the text published on YouTube's site in 2005: 'Easily share your videos with family, friends, or co-workers' (19 August 2005). Online photos and videos are not exactly tangible, but, given that they have offline equivalents, they are more so than objects of sharing such as thoughts, opinions, advice and ideas. Yet these too are reasonably concrete, and if we are asked to 'share thoughts with [our] friends' (Xanga.com, 4 December 2003) then it is pretty clear what is expected of us.

However, this is not the case with the new usage of sharing that characterizes SNSs. Particularly notable here are instances where users are urged to share their 'life', their 'world' or their 'real you', a term that appeared on the front page of Bebo at least up until May 2012. For instance, when, in 2007, LiveJournal says that it 'lets you express yourself, share your life, and connect with friends online' (25 April 2007), the object of sharing is *fuzzy* in that it is not obvious what sharing your life actually entails. This is also true of the phrase 'share your world', which appeared on the front page of Microsoft's Windows Live website (13 July 2011).

Significantly for my argument that we have here a new meaning of sharing, *the terms 'share your world' and 'share your life' do not appear before 2007* on any of the sampled websites. The idea of sharing your world is quite dense: on the one hand, to share your world with others is to tell them everything that is going on with you – what you are doing, thinking and so on. This draws on the sense of sharing as communication. However, sharing your world also includes uploading your photos to photo-sharing websites. Thus, Fotolog used to run with the tagline 'Share your world with the world' (23 January 2007) and Flickr used to exhort users to 'Share your life in photos'. Sharing your world or life would therefore appear to include a variety of communicative and distributive mechanisms. However, the rhetoric of sharing your world, and particularly that of sharing your life, also implies that you should not be alone: sharing your life is the opposite of living your life in isolation. I must share my life because it is distinct from your life, and the assumption is that you cannot know about my life unless I share it with you. Moreover, the way to share my life, according to the Windows Live site, is to 'Stay in touch', and this via 'Email, photos, movies, video, chat, and more', thus creating an association between sharing one's life and technologically mediated relationships.

In sum, the first objects of sharing in SNSs were concrete, and the use of the word 'sharing' here drew on familiar talk of file sharing (both in the sense of making your files accessible to others, and in the sense of distributing them). While sharing still has concrete objects today, these have been joined by fuzzy objects of sharing, which vastly extends the scope of what we are expected to share to include our lives or our worlds. In this context, sharing is more about communication than distribution, and is at one and the same time both much vaguer and far more inclusive.

No object of sharing

If the previous characteristic showed the emergence of the use of the notion of sharing with a fuzzy object, this one demonstrates the use of sharing without an object at all. This is significant for two main reasons. First, it reflects an assumption that users do not need to be told what to share, that the word is quite understandable without an object. Second, even more than when sharing is used with a fuzzy object, the total lack of an object gives the word a certain density. Taken together, this brings us closer to the understanding of sharing as the mode of participation in SNSs. Perhaps the clearest example of this social media sense of sharing is provided by the front page of Facebook, where we are told that 'Facebook helps you connect and share with the people in your life'. Likewise, in May 2012, MySpace's front page said that we can '1. Follow, 2. Get the latest, 3. Share' without saying exactly, or even fuzzily, what it is that we are to share.

This kind of usage of the notion of sharing *does not appear before 2005*, which suggests that only then did SNSs assume their users would be familiar enough with the idea of sharing to use it without an object: we no longer needed to be told *what* to share. In other words, in the examples just cited, the word 'share' serves as shorthand for 'participate in this site', while covering the range of possible activities in such sites – updating statuses, uploading photos, commenting on others' statuses, recommending links and so on.

If the use of the notion of sharing without an object reflects a belief among SNSs that users know what sharing is – that is, that a new meaning has taken root – it also enables multiple readings of the word. This was hinted at above in relation to the phrases 'share your world' and 'share your life'. However, when the word 'share' appears by itself, with not even a fuzzy object, then it is even more striking and the new meaning of sharing discussed here is even clearer. When the word 'share' appears by itself, its meaning is both clear and yet very dense. It is clear in that we know exactly what it refers to: the use of new ICTs, especially those through which we let other people know what we are doing, thinking or feeling, or – and these are usually the same technologies or platforms – through which we recommend websites and video clips to our friends. In saying that this meaning is dense, I mean that it includes a very wide range of practices: status updating, photo sharing, reviewing a book on Amazon, tweeting and so on. Finally, it is worth noting that this sense of sharing extends the communicative turn mentioned above in relation to fuzzy objects of sharing, though not at the expense of its distributive aspects. When enjoined to 'Share!', the logics of sharing include both telling people things and giving people things, though with the digital twist noted above that it is not a type of giving that depletes one's stock.

Wasn't sharing, now it is

The final aspect of the social media sense of sharing is that it has come to incorporate practices that existed a decade or more ago but that were not then called sharing. The widespread adoption of the term – particularly during 2005–7 – thus suggests that those using it seek to harness more than just its technical meaning of certain aspects of computer-mediated communication.

This point can be made through examples taken from the same website at different points in time. In 2005, for instance, Bebo's front page listed thirteen different things that members of the SNS could do there, including 'Write and Draw on other peoples' [*sic*] White Boards' and 'Keep in contact with friends at other Universities' (17 May 2005). However, in 2007, Bebo's front page was redesigned to include the text: 'Bebo is a social media network where friends share their lives and explore great entertainment' (14 November 2007). Furthermore, in 2009, a newly added graphic suggests that we 'Invite Friends to Share the Experience' (19 October 2009). So while the functionality of Bebo did not significantly change between 2005 and 2009, the way it presented that functionality did, in particular by adopting a rhetoric of fuzzy objects of sharing.

This is a process that can be seen in other sites as well. For instance, in 2002, the front page of the photo-sharing site Fotolog contained the text 'Make it easy for friends/family to see what's up with you' (5 June 2002). In 2007, though, it introduced the tag line 'Share your world with the world'. When the blogging SNS LiveJournal launched in 1999, it invited users to 'come and create your very own LiveJournal. Let the world know the story of your life, as it happens!' (27 November 1999). This is very much the same idea that is expressed through the phrase 'Share your life', and indeed, in 2006, the front page of LiveJournal included the text: 'Share your thoughts and photos with your friends' (1 March 2006). By 2007, the shift to the social media meaning of sharing was complete, as the site declared, 'LiveJournal lets you express yourself, share your life, and connect with friends online' (25 April 2007). This, then, is a clear example of how a single activity is represented differently at different points in time: 'Letting the world know' has become 'sharing'.

Why 'Sharing'?

Having outlined the meaning of sharing in the context of social media, I offer three answers to the question of why 'sharing' has become the term to describe participation in social media.

First, the notion of sharing is tightly interwoven with the history of electronic computing, from time sharing through to file sharing. As such, it is a term that was known by and very much available to developers of SNSs, who were certainly not the first to talk about transfers of data and information in terms of sharing.

Second, the notion of sharing, as it occurs in the context of social media, is extremely versatile. More specifically, if the 'traditional' definitions of sharing can be crudely divided into those in which sharing is communication and those in which sharing is distribution, now the concept of sharing incorporates both communication *and* distribution, especially when it is used with a fuzzy object, or with no object at all. Sharing on SNSs involves the *distribution* of digital content, in the form of links, photos, video clips and more. In this sense, I share something by letting someone else have it as well. Yet sharing on SNSs is also, and importantly, about *communication*, particularly through the practice of updating one's status. Here, sharing is telling. At least some of what we are encouraged to share on SNSs is our feelings, and so there is an overlap between a common spoken use of the term and its social media meaning. However, letting people know your opinion of current events, your location, or any of the minutiae of your day-to-day life is, in social media, also called sharing. Another way of thinking about this is to suggest that 'sharing' in the context of social media covers both the transmission and ritual conceptions of communication, with their respective emphases on the 'extension of messages' and 'fellowship and commonality' (Carey, 1989: 18).[24] With the rise of sharing on social media we have another example of 'industrially-driven, technologically-mediated message machines (the press, cinema, and broadcasting) [...] shar[ing] semantic quarters with the venerable arts of interpersonal talk' (Peters, 1994: 117).

The activity of sharing in social media is thus remarkably broad – far broader than any of the other words that might be found in SNSs' earlier self-descriptions, such as 'express', 'connect', 'post', 'blog' or 'socialize'. For sites that want you to distribute photos *and* communicate your emotions, the notion of sharing covers all bases.

The third part of the explanation for the spread of the notion of sharing lies in its positive connotations of equality, selflessness and giving, in combination with its resonance with what is viewed as the proper mode of communication between intimates. In brief, sharing is associated with positive social relations, as expressed through the popular phrase 'sharing and caring', which has been appropriated by SNSs so as to infuse their services with the positive implications of that term. This idea is well exemplified through a reading of Facebook's blogs about itself and developments in the services it offers. For instance, in a blog entry from 2009, we are told that

> [t]he Share button enables you to take content from across the Web and share it with your friends on Facebook, where it can be re-shared over and over so the best and most interesting items get noticed by the people you care about.[25]

Disregarding the fact that if the content you have shared is 're-shared over and over' it is unlikely that you will even know the people who are noticing it, let alone care about them, the connotation of this quote is quite clear: your sharing is an expression of your caring. A similar rhetorical move was made by Yahoo! on the front page of its Pulse network, which included the text: 'Share what's important to you with the people you care about' (19 July 2011). Windows Live Spaces presented itself similarly in 2006: 'Windows Live Spaces is a free online software service where you can blog, share pictures and connect with the people you care about' (2 September 2006). Finally, on Bebo's About page, we learn that 'Bebo is a popular social networking site which connects you to everyone and everything you care about.'[26]

Of course, it is not only SNSs or social media enterprises more generally that have harnessed the positive connotations of sharing. For instance, in a British ad campaign from 2011, mobile telephony company T-Mobile ran with the slogan 'Life's for Sharing'. One ad, for example, tells us that 'Some things in life you just have to share', followed by an offer of free minutes. This suggests that the idea of sharing has grown in popularity in spheres beyond the internet and has become a useful term for marketing purposes.

Sharing on social media, then, is a concept that incorporates a wide range of distributive and communicative practices, while also carrying a set of positive connotations to do with our relations with others and a more just allocation of resources. This, of course, is not to say that people participate in SNSs as an expression of their care for the people with whom they are sharing, but it is to say something about why this word, and not any of the others mentioned above, has become the *sine qua non* in the self-presentation and the praxis of SNSs.

Sharing and Mystification

In this section, I propose to focus on the rhetorical power of the notion of sharing and to show how it serves to paper over the commercial aspects of the ways in which many SNSs operate. Put differently, I wish to highlight the seeming inconsistencies, not to say contradictions, between the rhetoric of sharing as described above, and the actual practices of SNSs.

While the critiques of social media are many and varied, of most relevance here are those that deal with the ways in which SNSs and other social media enterprises make money, or those that explain how companies use social media tools and platforms to exploit the 'free labour' (Terranova, 2000) of the users of and visitors to these platforms (see also Fisher, 2015). This critique has two distinct targets. The first is the use of people's free labour to perform tasks that the company would otherwise pay people to carry out. This is the thrust of van Dijck and Nieborg's (2009) critical analysis of Web 2.0 manifestos, which attacks the tendency of companies to crowdsource certain tasks (and see also Morell, 2011). The second target of critiques of social media is the way that they monetize their users' activities. That is, Facebook makes money not by asking its users explicitly to perform tasks for Facebook, but rather by aggregating and selling the data produced by the members' interactions with one another on the site and, through the Like button and Facebook's 'frictionless sharing', with other sites. What these and other critiques throw light on is the way that, through increasingly sophisticated techniques of data mining, SNSs are able to sell website real estate to advertisers based on the promise of targeted advertising at an unprecedented resolution (Zarsky, 2002). These ideas, which have been formalized by researchers (see, for instance, Fuchs, 2011; Zimmer, 2008), were succinctly if pithily expressed by a user of MetaFilter, a weblog community: 'If you're not paying for something, you're not the customer; you're the product being sold.'[27]

In brief, there are two ways in which the use of a rhetoric of sharing on Facebook (and elsewhere) serves to mystify the commercial relations just mentioned. The first is the idea that the more we share (note: no object), the better the world will be. This can clearly be seen in the letter Zuckerberg attached to Facebook's IPO (initial public offering) filing in January 2012.[28] There, he writes that 'Facebook was not originally created to be a company. It was built to accomplish a social mission – to make the world more open and connected', and that Facebook's objective is to 'strengthen how people relate to each other'. Relationships, continues Zuckerberg, 'are how we discover new ideas, understand our world and ultimately derive long-term happiness.' Moreover, '[p]eople sharing more – even if just with their close friends or families – creates a more open culture and leads to a better understanding of the lives and perspectives of others.' Here, then, sharing is represented as a mechanism for improving human relations and making the world a better place.[29] (For an analysis of this and other letters attached to high-tech IPOs in which the CEOs claim they are not in it for the money, see Dror (2013).) I shall leave the critique of the use of 'sharing' by Facebook to Frick and Oberprantacher, who bemoan the fact that 'the very institution which has defined *connecting* and *sharing* its unique selling proposition is itself a profit-oriented enterprise that is neither reflecting the idea of *the commons*, nor is it up to public scrutiny' (Frick and Oberprantacher, 2011: 22; emphasis in original).

The second mechanism of mystification involves the way in which Facebook's relationships with advertisers are described in terms of sharing. The following quote expresses this well, whereby Facebook seeks to assure users that 'we don't share information we receive about you with others unless we have received your permission'.[30] Regardless of the fact that the quotation is expressed in the negative, the relationship between Facebook and its advertisers is represented in terms of sharing; that is, the transmission, or selling, of data by Facebook to advertisers is described as sharing. Facebook is most certainly not alone here. Under the heading, 'Information we share', Google's privacy policy (in July 2016) included the statement: 'We will share personal information with companies, organizations or individuals outside of Google when we have your consent to do so.'[31]

It is possible that high-tech companies are using the word 'sharing' in a neutral way, similarly to time sharing in the 1950s. However, when Facebook, Google and others talk about sharing information about us with third parties, rather than selling it, or even just transferring or sending it – two much more neutral terms – a parallel is created between our online social interactions with our friends on the one hand, and commercial interactions involving some of the wealthiest organizations on the planet on the other. This is not to say that this is an intentional move on the part of the SNSs. As already mentioned, the word 'sharing' would have been accessible to them from its pre-existing associations with computing, as well as from the popular meaning of sharing as talking about one's feelings. However, it is still the case that every time we share something online, we create traces of data, which constitute the hard currency of commercial social media organizations. The more we share online, the richer the online platforms we are sharing on become.

Conclusion

This chapter has presented the rise and rise of 'sharing' in social media, offering an analysis of what is, for all intents and purposes, the constitutive activity of social media. It has shown that 'sharing' has become the word of choice to describe the values of the internet at large, and the way in which we participate in social media in particular. We have seen how the word builds on more 'traditional' meanings of sharing, enfolding within it both distribution and communication as well as the usage of sharing in the context of computing. The data show that the years 2005–7 constitute a watershed in terms of the use of the concept of sharing. Terms such as 'share your world' or 'share your life' did not appear before then; similarly, the injunction to share (without any object at all) did not appear until the second half of the 2000s. I also showed how certain activities, such as keeping in touch, over time came to be described as sharing. Finally, I offered a critique of how the notion of sharing as deployed by Facebook serves its self-representation as leading to a better world, as well as helping to mystify its commercial relations with advertisers.

The rise of sharing in social media, or the way that 'sharing' became their constitutive activity, is understood here in relation to a popular conception of the internet as democratic, open and free, and as a platform for non-hierarchical communication. Turner and others have shown us how the internet was constructed as prosocial; the notion of 'sharing' came later. This idea was captured by a Twitter user: 'The internet has always been about sharing, "Sharing" is not a new thing, it's a catch phrase but the core essence has always been, sharing.'[32] Similarly to the quotes mentioned at the start of this chapter, here too we read the 'internet has always been about sharing'. This tweet notes that 'sharing' may be a 'catchphrase', but that sharing has always been at the 'core essence' of the internet. 'Sharing', then, is both a kind of

Platonic form of which the internet partakes *and* a catchphrase – as indicated by the quotation marks round the second instance of the word in the tweet. Catchphrases come and go; core essences do not. In this chapter, I have shown that 'sharing' has enjoyed an ascendancy that might make it seem like a catchphrase, or buzzword. I have also shown that there is something about how we understand the internet today that resonates with the values we associate with sharing. However, this is not because of the internet's 'core essence', but is rather the function of discursive work carried out by, among others, its early users.

Today it certainly feels as if the 'core essence' of the internet is 'sharing'. This is because, as demonstrated above, the term has come to cover the gamut of online activities, while the internet as an abstract whole is seen by many as responsible for specific positive processes, such as the Arab Spring. If 'sharing' is a central metaphor in contemporary society, its role as the constitutive activity of social media, and perhaps the internet at large, is a crucial part of that.

4
Sharing Economies

Over the last few years, the term 'sharing economy' has truly entered the mainstream, where it has become predominantly associated with companies such as Airbnb and Uber. While not wishing to overstate the significance of online searches, Figure 4.1 from Google Trends shows a spurt in American internet users' interest in the sharing economy that is pretty hard to ignore. The rise and rise of the sharing economy (or what some call the 'so-called sharing economy') has attracted a great deal of critical attention (see especially Scholz, 2016; and Slee, 2015). One often-heard critique is that the sharing economy has nothing to do with sharing. Airbnb is not a platform for sharing, but rather for short-term renting, we are told. Uber is not a platform for car sharing, say the doubters, but rather an unregulated and exploitative system whereby people work without the faintest hint of employee protection or social rights.[1] If this is the sharing economy, then its message is pure neoliberal ideology: if you have any spare resources (including your time, that drill you are not using right now, your spare bedroom) and you are not monetizing them, then do not complain about being poor. There is a part of the sharing economy that turns us all into micro-entrepreneurs,[2] and that looks mercilessly upon those who refuse to participate. This is clearly an important development in high-tech capitalist societies and one that is receiving the attention it requires and deserves.[3]

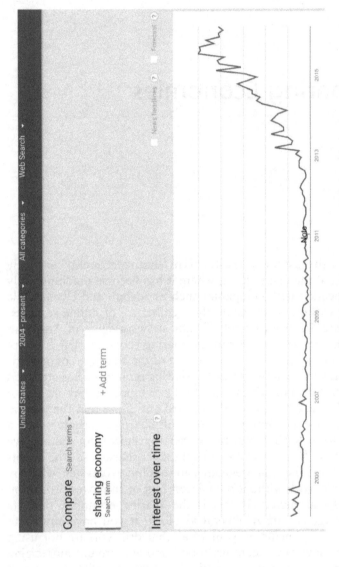

Figure 4.1 Google Trends graph for 'sharing economy', 2005 through August 2015

In this chapter, though, I would like to expand the view of the sharing economy to include enterprises that people might more willingly accept are 'really' about sharing, and I would like to issue a reminder that the term also has an application in the context of production, and not only consumption. I will also look at the digital nature of the sharing economy and make an argument about the term 'sharing' and the public/private divide. Primarily, this chapter tries to understand what the word 'sharing' *does* in the context of the sharing economy (whatever that is). My objective is not to engage in the argument as to whether the sharing economy really does involve sharing (though I do, just a little), but I do hope to show, at the very least, that the term is not merely a manipulative marketing ploy. In order to do this, we shall need to look into the short history of the term 'sharing economy'.

The Sharing Economy: Since When?

A central aspect of the sharing economy – and of the notion of sharing in general – is the way it faces the past. As we shall see, much of the discourse around the sharing economy involves positioning it in relation to a prelapsarian past, a romanticized view of how we, members of capitalist, consumerist societies, used to live; the kind of view of the past that Walter Benjamin called an ur-past (Benjamin and Tiedemann, 1999). Another position is that we have never stopped sharing, but that new technologies are giving these practices a boost: thus, the tagline of the collaborativeconsumption.com website is 'Sharing reinvented through technology'.[4] A central concern of this book is when, historically speaking, the word 'sharing' came to be adopted to describe various practices. Given this, I am interested not only in when the sharing economy as we know it today took off, but also when it started to be called the 'sharing economy'.

The term 'sharing economy' is actually quite a new one. Recent interest in it is suggested by the Google Trends graph above (Figure 4.1). Another Google tool – its Ngrams offering – enables us to explore not when people started *searching* for information about the sharing economy, but rather when people started *writing* about it. Based on the Google project of scanning and performing character recognition on millions of books, we can search these books for words and terms and thus learn when they became popular. Figure 4.2 shows the Ngram for the term 'sharing economy' from 1940[5] through to 2008 (the most recent date available).

Tracing the line from left to right, the first thing we notice is a little blip in the 1950s. However, the references here to a 'sharing economy' are actually for the term 'profit-sharing economy', which is also picked up by this search string. The same is true of the surge from the mid-1970s through to the end of the 1980s, when economists were trying to establish whether incentivizing workers by giving them a share of the company's profits improved their productivity. So not only does the spike in the 1980s not tell us anything about today's sharing economy, it actually reflects increased writing about profit-sharing mechanisms. Of more interest is the rise in writing (in books scanned by Google) about the 'sharing economy' from around 2003.

Even here, though, the first reference to the 'sharing economy' as we understand the term today is actually from 2007. In the book *Innovation Nation* (Kao, 2007), the sharing economy is mentioned in passing, and without definition, in relation to the Creative Commons licence spearheaded by Lawrence Lessig. In a book about the future, *Get There Early* (Johansen and Johansen, 2007), Howard Rheingold is credited with talking about the sharing economy, but no reference is given for that (and the reference supplied does not include Rheingold talking about the sharing economy;[6] in Rheingold's (2007) *Smart Mobs*, a book in which sharing is discussed a fair amount, the term 'sharing economy' does not appear; indeed, we have already seen, in Chapter 3, that Rheingold does not talk much about 'sharing'). Later, in *Get There Early*, the term 'sharing economy' appears again, in quotation marks, and in relation to the workplace. Here Yochai Benkler's work on peer production is referenced, which is undoubtedly of great relevance, but Benkler himself does not use the term 'sharing economy', neither in his article, 'Sharing Nicely' (Benkler, 2004), nor in his influential book, *The Wealth of Networks* (Benkler, 2006).

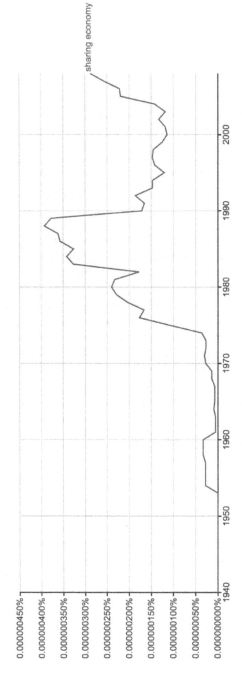

Figure 4.2 Google Ngram for 'sharing economy', 1940–2008

Nonetheless, the appearance of the term in a book from 2007 suggests that it was around then that its current sense was starting to reach popular consciousness. Indeed, the major publication that brought the term 'sharing economy' to a wider audience was *Remix* by Lawrence Lessig (2008). There, inspired by Benkler, Lessig discusses 'a commercial economy, a sharing economy, and a hybrid of the two' (p. 118). For Lessig, 'Of all the possible terms of exchange within a sharing economy, the single term that isn't appropriate is money' (p. 118). This is a position that he illustrates by telling us about the time he offered to pay the teen sitting next to him on a plane $5 to rent one of his DVDs, or about the time his neighbour tried to give him $5 after he had helped the neighbour jump-start his car. In Lessig's discussion, the 'sharing economy' refers to all interactions that do not involve money, including marriage, playing softball, friendships and the like. His paradigm of the sharing economy, though, is Wikipedia, because 'Its contributors are motivated not by money, but by the fun or joy in what they do' (p. 162).

Other instances of the sharing economy that interest Lessig include Linux, produced by 'thousands volunteering to write code' (p. 163); Project Gutenberg, whose collection of freely accessible texts has been produced by volunteers; Distributed Proofreaders, whereby volunteers proofread the digital versions of the books on Project Gutenberg, and distributed-computing projects more generally, such as SETI@Home, and Einstein@Home; the Internet Archive; and more. It is notable that these are all examples that focus on *production* and emphasize the role played by volunteers in producing value for others.

It is also notable, of course, that for Lessig the broader context of the sharing economy is that of culture, and especially music (to which we shall return in Chapter 6 on file sharing). This emerges more strongly from a correspondence I initiated with Lessig, in which I asked him where he found the term 'sharing economy'. He referred me to Joichi Ito of the MIT Media Lab, who in turn reported that he first came across it in an article from 2004 in the erstwhile *Business 2.0* magazine.[7] That article describes the emergence of a 'sharing economy' in the music industry and the spreading use of Creative Commons licences (which were written by Lawrence

Lessig, who is interviewed in that piece...). The 'sharing economy' tag for blog posts on Ito's blog also dates back to 2004.[8]

Talk of the sharing economy, then, was first associated with the music industry and the technological challenges posed to it by the new modes of distribution offered by the internet (including the ability to easily reproduce and distribute MP3 files). In this regard, it was an economy that was associated with new technologies, seen as technologies of sharing (see Chapter 3), and with the absence of money, or even a subversive approach to regular capitalist relations.

Notably, then, the term has always referred to both production and consumption, with the spheres of open source software (production) and file sharing (consumption) providing important initial sites identified as instantiating the 'sharing economy'. Also, the term emerged between 2005 and 2010, thus following the rapid spread of the word 'sharing' in the context of social media, pre-empting the argument to be made below that the sharing economy is digital, and that it is the word 'sharing' that does much of the work in discursively constructing it as such.

The Sharing Economy, Early Childhood and Ancient History

One way that the sharing economy is promoted is by describing it as a return to prelapsarian modes of behaviour. The narrative offered by the sharing economy, or at least certain of its representatives, is this: we used to be good, we went bad, sharing will make us good again. This is an extremely simple message and a powerful cultural trope, perhaps as old as the story of the expulsion from Eden. This narrative operates on two levels – the individual and the social. In both instances sharing is depicted as a natural and innate behaviour that is bashed out of us by the all-powerful and corrupting forces of capitalism. The Black Friday stampede at a Walmart in Nassau County, New York, in November 2008, in which temporary worker Jdimytai Damour was trampled to death by hundreds of bargain hunters who had broken down the shop's doors, is held up as precisely the kind of unnatural behaviour that capitalism has led us to (Botsman and Rogers, 2010).

The rediscovery of sharing is, on this account, a return to nature, a return to how we are meant to be, both as individuals and societies. This promise of a return to an uncorrupted, natural state of being – a prelapsarian state of affairs – is an important part of the work that the word 'sharing' does in the 'sharing economy'. This, we might say, is what gives it its 'specific *evaluative accent*' (Voloshinov, 1986: 103; emphasis in original). It reinforces an association with childhood, and it suggests the possibility of a society in which resources are distributed fairly and people live in harmony with nature rather than exploiting it and one another.

This kind of past is what Walter Benjamin called an 'ur-past'. As Susan Buck-Morss describes it, this is a 'mythic time when human beings were reconciled with the natural world' (Buck-Morss, 1989: 114). As put by Benjamin himself in *The Arcades Project*:

> The utopian images that accompany the emergence of the new always concurrently reach back to the ur-past.[9] In the dream in which every epoch sees in images before its eyes the one that follows it, the images appear wedded to elements of ur-history. (Cited in Buck-Morss, 1989: 116)

Sharing, childhood and human nature

We readily associate sharing with childhood. Conduct an image search, if you would, for 'sharing' in your search engine of choice and observe the results: children holding hands, children messily eating an ice cream together, children playing harmoniously. Or recall from Chapter 1 the promotional image used by the website ecoSharing.net of two girls, holding hands as they walk through a park, each with one bud of a single headphone set in one ear, listening together to music. The tagline, dripping with nostalgic yearning for a happy, carefree childhood, reads, 'Do you remember how much you enjoyed sharing what you owned?' (Figure 1.3). Putting aside memories we may also have of being forced to share a favourite toy with a not-so-favourite cousin, according to this conception sharing returns us to a carefree time of life.

As well as appeals to our emotions, such as that by ecoSharing.net, some proponents of the sharing economy turn to science to establish that we are natural-born sharers. For instance, in *What's Mine is Yours*, Botsman and Rogers (2010) refer to Michael Tomasello's (2009) best-seller, *Why We Cooperate*, in order to advance their position that sharing is what children do unprompted. Specifically, they cite studies in which very young children try to help adults to open doors, find things they have lost or pick up things they have dropped. Botsman and Rogers' gloss on Tomasello's work – and it is their take that is important here, not Tomasello's own presentation – is that 'Tomasello argues that empathy and cooperative behaviors are not learned from adults or done out of expectation of a reward. *Children are sociable and cooperative by nature*' (p. 69; emphasis added).

However, according to Tomasello (as described by Botsman and Rogers), 'human beings have a selfish streak', with cooperativeness laid on top of this. At this point Botsman and Rogers' account becomes slightly less coherent as they try to explain what has gone wrong with us: 'For the past fifty years, children have grown up in a hyper-individualistic society, so it is not surprising that their inherently selfish side has overshadowed their equally natural social inclination to share' (p. 69).[10] So now, confusingly, children are 'inherently selfish' as well as being 'social and cooperative by nature'. Ultimately, though, the representation of childhood promoted by Botsman and Rogers is that our inclination to share is *presocial*, and that this inclination is overcome by the hyper-individualist society in which we have been living for the past fifty years. The implication is that by sharing we shall return to who we really are meant to be.

The same point about sharing is made by the Innovation Director at Collaborative Consumption, Lauren Anderson. In an interview she gave in 2011 to TechCrunch,[11] she said that 'it is innate in us to want to share and collaborate'. This kind of approach to sharing – that it is an innate, natural inclination that is bullied out of us as we grow older by a materialist, individualist society – suggests that sharing-economy proponents have in mind the cultural model of the 'innocent child'. In their seminal book, *Theorizing Childhood*, James, Jenks and Prout (1998) present different models of childhood which, they argue, inform how we think about children. The two main models are the 'evil child' and the 'innocent child'. While the former is the child that requires discipline and institutional restraint, the latter is pure of heart and exudes natural goodness. While society at large tends to 'vacillate between the child as innocent and original sinner' (Critcher, 2008: 101), for proponents of the sharing economy there would seem to be little deliberation. If childhood itself is 'often invested with our most intense fantasies and fears' (Buckingham, 2006: 76), then the fantasy of childhood evoked by writers such as Botsman and Rogers is one of kind and considerate children who are spoiled by the cruel society in which they (we) grow up.

If this applies to us as individuals, a similar kind of argument – that sharing is what comes naturally – is put forward by Yochai Benkler in his paean to cooperation, *The Penguin and the Leviathan* (2011), in relation to humans as a species. There, he discusses the centuries-long debates over human nature, represented by Thomas Hobbes and Adam Smith on the one hand, and David Hume and Jean-Jacques Rousseau on the other. In his arguments for cooperation and collaboration, Benkler mentions a study which shows that 'our brains light up differently when we cooperate with other humans; it makes us, at least many of us, happier' (Benkler, 2011: 236). Botsman and Rogers also refer to a neurological study that asked participants whether they would donate a sum of money to charity or keep it for themselves. Apparently, the 'part of the brain associated with bonding and attachment lit up for those who chose to give the money away' (Botsman and Rogers, 2010: 132).

Botsman and Rogers also bring up the work of primatologist Frans de Waal, who is known for arguing that humans, like much of the animal kingdom, are naturally empathetic (see de Waal, 2009). They tell of a study whereby capuchin monkeys would happily trade pebbles for slices of cucumber, until they saw another capuchin monkey receiving a grape in return for a pebble. At this point, their innate sense of fairness kicked in, and the monkeys receiving cucumbers stopped trading. According to Botsman and Rogers, this was because they had learned what the fair trade for a pebble was.

As I understand them, the point of these examples is to show that human beings are naturally empathetic, caring and sharing beings. However, their logic is not always entirely clear: if we are hardwired for cooperation (that is, if cooperation is presocial, innate, not learned), and if sharing gives us such pleasure, then how is it that a mere 100 years of consumer society (or fifty years of 'hyper-individualism', to use Botsman and Rogers' turn of phrase) have managed to overcome millions of years of biological evolutionary development? We might equally wonder whether the fact that monkeys (who are taken as representing our evolutionary forefathers) cooperate means that we ought to. (In fact, there are very many examples of is-ought leaps in this literature.) So this is another instance where it is not clear quite how 'innate' sharing is being seen when Botsman and Rogers talk about 'innate behaviours that make it fun and second nature to share' (Botsman and Rogers, 2010: xx). It is not my intention to quibble here, but this short quote seems to encapsulate the confusion between the innateness of sharing and its being something that can become second nature to us. My point being, of course, that if it is innate it is *first* nature. However, it is not actually important to me here whether the science on which the sharing economy is based is 'right' or 'wrong'. What interests me is how the research is co-opted by proponents of sharing and how we might understand that.

Sharing and 'primitive' human societies[12]

An extremely common trope among supporters of the sharing economy is that by sharing we are cleverly exploiting technology in order to return to a social way of being that characterized our ancestors. The idea that 'primitive' human societies can re-teach us how to live dates back at least to the eighteenth century (see Kozinets' (2002: 29) reference to Weiner (1992)). Nor is it particularly new among media researchers. Richard Coyne, for instance, mentions McLuhan's belief in a 'technoromantic return to tribal society, in which commerce was dependent on kinship' (Coyne, 2007: 101).

Thus, Botsman and Rogers can write that 'consumers are yearning to go back to a time when markets meant community-based, traditional relationships with strong ties' (p. 51). But more than that, when we share, they say, we are imitating our 'Paleolithic ancestors in the stone age', who lived in small groups and hunted in packs. (Another somewhat Kropotkin-esque (1902) parallel presented by Botsman and Rogers is that of dolphins, which also hunt together.) 'Following a kill,' they write, 'the meat was cut into pieces and shared with everyone in the camp' (p. 68). Moreover, Botsman and Rogers tell us that 'Anthropologists believe that this mutualism (people helping each other) and reciprocity ("I'll give you meat today, you give me meat in the future") are hardwired human behaviours that serve as the basis for human coopera-tion and are the core of our experience' (p. 68). In an equally sweeping statement, author and activist David Bollier asserts that 'evolutionary scientists make it clear that the human species has survived and evolved only because it has had great capacities to cooperate and collaborate and to build stable communities based on trust and reciprocity' (Bollier, 2010: 13). This would seem to be common wisdom across all wings of the sharing economy, namely, that sharing is, and has always been, 'necessary for our survival as a species' (Harris and Gorenflo, 2012: xii).

References to the 'tribal' are also quite widespread in the context of the sharing economy. For instance, futurist Mark Pesce says that 'across the rest of this century our shared future will look more like our ancient, tribal past than our recent history' (Pesce, 2012).[13] Environmentalist and journal-ist Leo Hickman said that the sharing economy 'has been around for ages' (Hickman, 2011). Relatedly, an article in *Marketing* magazine suggests that 'we've always done this. Charles Darwin and, latterly, Richard Dawkins argued that collaboration represented an evolutionary advantage, and humans seem to be natural collaborators' (Walmsley, 2011). As Walmsley continues, through the sharing economy 'we just re-adopted an old behaviour that, once again, became useful'. Similarly, Lisa Gansky, in her influential book *The Mesh: Why the Future of Business is Sharing* (Gansky, 2010), offers several reasons as to why the emerging sharing economy is like nature itself. For instance, it 'abhors a vacuum', it is 'resilient and adaptive' and it uses waste efficiently. Once again, sharing is presented as a presocial, desirable state of nature.

Sharing, then, is constructed as the state of nature of human beings at two levels: the individual and the social. At the individual level, we have seen how sharing economy advocates describe children as natural sharers, implying that we are all born innocent and good before being sullied by individualist, materialist society. At the social level, sharing economy advocates describe human societies as naturally sharing, implying again that the natural state of affairs is one of cooperation, trust and togetherness. They do this by referring to prehistoric societies and, à la Kropotkin and his concept of mutual aid (Kropotkin, 1902), to animal behaviours. Once more the message is that we have ruined a good thing, and that sharing will help us regain a paradise lost.[14] These are important constituents of the metaphor of sharing. Of course, given the colonialist view of 'primitives' as childlike – or the notion that 'ontogeny recapitulates phylogeny' – the ideas of redeeming a corrupted self and redeeming a corrupted society are actually quite closely related.

The Sharing Economy and Network Technologies

The sharing economy, or what most people mean by the sharing economy today, is digital. (One could doubtless think of examples of sharing that are not digital, but they are not what is intended by the currently accepted meaning of the term 'sharing economy'.) Access to the sharing economy is mediated by websites and apps. I have already established that the mode of participation in social media and network society more generally is sharing. However, these two semantic fields of sharing – the sharing economy and social media – are linked not only by the homophone of 'sharing', but also, and more importantly, by the constellation of values and associations the word ignites.

In this section I shall present some of the ways in which the sharing economy – and particularly collaborative consumption – is discursively constructed as technological, pointing to the role attributed to technology as enabler, and looking at some of the metaphors deployed when the sharing economy is discussed. I shall then discuss the idea that technology is a driver of the sharing economy. In particular, I shall examine the claim commonly put forward in both academic and popular sources, that today's younger generation – known by such terms as the 'Millennials', or 'Generation Y' – have been somehow primed to share on account of their deep familiarity with online sharing. My argument will be that the word 'sharing' is doing much more work than people seem to give it credit for.

Technology as enabler

Commentators who view technology as an enabler of the sharing economy tend to draw on one of two narratives: first, that technology is enabling processes that are being driven by financial and environmental concerns; and, second, that technology is enabling processes that express behaviours that are described as profoundly human in the manner outlined above.

The most neutral views of the role of technology in the growth of the sharing economy merely note that social media and computer networks are being used by consumers for sharing purposes. 'Consumers tend to be leveraging social media to create communities around collaborative consumption', writes Katie Cincotta (2011), quoting Booklending.com founder Catherine MacDonald, while Kevin Courtney (2011) notes that, 'All over the globe, people are using social networks and other web platforms to trade, swap, rent or barter goods, skills, services or expertise'.

Other commentators, however, go beyond merely noting that collaborative consumption is often carried out by means of social media, giving it a more active role as an 'enabler'. In an article in *The Times*, for instance, journalist Alexandra Frean (2010) writes that 'The rise of collaborative consumption [...] has been made possible by the internet, social networks, mobile devices and location-based GPS services enabling the ready exchange of data concerning location, availability, price, access and so on', while a feature in the *Guardian* describes 'the internet and social networking' as 'lubricants for collaborative consumption' (Hickman, 2011). Echoing the metaphor of lubrication, an article in the *Sydney Morning Herald* about car sharing explains that

> Scheduling a fleet of a few hundred cars across several thousand drivers once required vast and complicated record keeping, greased with plenty of labour. That same capability, computerised, networked and delivered into every home and onto every mobile, drops the bar low enough that the previously impossible becomes realisable. The friction associated with this so-called 'collaborative consumption' has been removed. (Pesce, 2012: 11)

However, as well as being a friction-reducing lubrication, technology has, according to some accounts, served as a catalyst to processes that were already under way. For example, Mireya Navarro (2010: 22) writes that 'Social media like Facebook lend momentum to [collaborative consumption] as people join forces to trade, share or negotiate better deals from retailers.' Or, as put more forcefully by Leo Hickman (2011: 10), 'the arrival of internet-enabled social networking, coupled with "geo-located" smart phones, has super-charged a concept that was already rapidly gaining primacy owing to the twin pressures of our environmental and economic crises.' Here, then, people are already consuming collaboratively for reasons that have nothing to do with the technological landscape and are everything to do with economics and the environment. Technology is not a necessary condition for these processes; rather, it is presented as 'super-charging' them and enabling them to reach unprecedented heights.

There is a sense, then, that technologically enabled sharing is somehow a return to an older and better way of living. As one article argues, 'Real-time technologies are starting to be used to create similar virtual community bulletin boards akin to the "good old days" when everyone knew each other' (n/a, 2010: 30). Similarly, Jenna Wortham (2010) describes collaborative consumption as 'a throwback to the good old days when people actually spent time socializing at local markets'. Sharing, according to these and other excerpts, is an old social practice that technology is enabling in new ways. This can be added to what was said earlier about sharing being an age-old practice; here, though, rather than looking back millennia, we are encouraged to look back to the way our grandparents used to live.

Technologies, then, are represented as enabling collaborative consumption in different ways: they are the media through which the sharing economy takes place; they are accelerating processes that were already under way; and they are allowing us to reinstate old behaviours ('old' as in how our grandparents used to live, and 'old' as in ancient and 'tribal'). In all of these instances, the motivations for engaging in the sharing economy – saving money and conserving the environment – lie beyond the technologies that are said to be enabling it. However, as we shall see, many newspaper articles present a view whereby social media and ICTs not only enable collaborative consumption, but are actually *drivers* of it.

Technological metaphors and terminology in the discursive construction of collaborative consumption

As well as explicit references to the relationship between technology and the sharing economy just mentioned, the metaphors and other terminology employed in popular texts about the sharing economy point to its construction as a technological phenomenon in another, more subtle way. I wish to make one main point in this regard concerning the use of the metaphor of P2P (peer-to-peer).

The term 'peer-to-peer' refers to a specific type of network configuration, namely one in which the various members are connected to one another directly, or through other members, but not through a central hub. As defined by Michel Bauwens of the P2P Foundation, it is 'any system which allows agents to freely and permissionlessly interact with each other' (Bauwens, 2011: 42). This is not to say that certain nodes are not more densely connected than others, but it is to say that the network lacks a centre, an organizing core. According to Bryn Loban (2004), 'The term Peer-to-Peer (P2P) appeared in the world media around the year 2001 with a music sharing system called Napster.'[15] However, he adds, 'P2P is, strictly speaking, nothing new, but is more of an evolution than a revolution', with Minar and Hedlund asserting that the 'original internet was fundamentally designed as a peer-to-peer system' (2001: 4). The term 'peer-to-peer' thus refers to computer networks, and to use it in relation to networks of people is to use it metaphorically.

The notion of peer-to-peer has a number of connotations that sit well with the overall agenda and outlook of the sharing economy. First, there is a parallel between the notion of redundancy in a peer-to-peer computing network – the idea that if one node falls, others can take up the slack – and the key concept of idling capacity in collaborative consumption, defined by Botsman in an article for the *Observer*, as 'the untapped social and economic value of under-utilised spaces, skills, time, gardens, and "stuff"' (Botsman, 2011: 8). Second, both types of peer-to-peer system are decentralized. For peer-to-peer loan schemes in particular, this aspect of the metaphor is especially attractive given the context of the fallout of the 2008 financial crisis and a mistrust of banking institutions. The peer-to-peer metaphor thus holds out the promise of unmediated contact between individuals, a promise that is all the more appealing when the traditional mediators are perceived as greedy and obstructive, and, post-2008, failing. Relatedly, the peer-to-peer metaphor implies an absence of hierarchy and a sense of egalitarianism, which resonate strongly with the concept of sharing as it is understood today.

Technology as driver

The way in which 'sharing' is most active as a metaphor in constructing the sharing economy as technological and in infusing it with digital connotations is found in the postulated causal connection between the sharing that young adults have done online, and their willingness to share stuff offline. Very briefly, the idea that is put forward is that having become so accustomed to and adept at online sharing, young adults are more open to sharing their stuff. This is a notion that is repeated frequently in all kinds of texts about the sharing economy. My objective here is not to evaluate the empirical accuracy of the claim, but rather to examine the role played in it by the word 'sharing', drawing on the theoretical foundations I laid down earlier, and particularly the distinction between sharing as communication and sharing as distribution.

Examples of the causal claim that sharing online leads to sharing offline are plentiful, and the claim itself seems quite central to Botsman and Rogers' book *What's Mine is Yours* (2010). The logic that Botsman and Rogers offer starts with online communities: 'The phenomenon of sharing via increasingly ubiquitous cyber peer-to-peer communities such as Linux, Wikipedia, Flickr, Digg and YouTube is by now a familiar story', as they argue that 'Collaborative Consumption is rooted in the technologies and behaviours of online social networks' (p. xx). Indeed, collaborative consumption, they say, started online, as evidenced by people 'posting comments and sharing files, code, photos, videos and knowledge' (p. xx). And then comes the causal argument, according to which 'we are starting to apply the same collaborative principles and sharing behaviours to other physical areas of our everyday lives' (p. xx).

This is an argument that Botsman is quite fond of, and she made it in her influential TEDxSydney talk of 2010.[16] In that presentation, one of the reasons she gives for the fact that 'things are changing' is to do with digital natives, or 'gen-Y', who, she says, 'are growing up sharing – files, video games, knowledge; it's second nature to them'. She then goes on to state: 'I genuinely believe we're at an inflection point where the sharing behaviors, through sites such as Flickr and Twitter, that are becoming second nature online, are being applied to offline areas of our everyday lives.' The point I wish to make here, and which shall be reinforced in the following paragraphs, is that the 'sharing behaviours' we carry out on Flickr and Twitter are *communicative*, while the sharing that we do offline in the sharing economy is *distributive*. The association between uploading photos or tweeting and participating in collaborative consumption is, on the face of it, unclear. This is not to say that there is not an association, but I fail to see the association made by Botsman, other than recognizing that in both instances the constitutive activity is called 'sharing'.

The same association is imputed in a *New York Times* article from 2009 about Zipcar (Levine, 2009: MM36), in which we read that Zipcar owes its success to an 'Internet-bred generation of urbanites who have come of age sharing information (Wikipedia), music (Napster), software (BitTorrent), videos (YouTube), entertainment (Netflix) and the excruciating minutiae of daily life (anyblog.com)'. Again, this kind of statement seems passable, but really it begs the question of why blogging should in any way prepare one for being a customer of Zipcar, and raises the question of whether the word 'sharing' itself is not conjoining two separate spheres of life. This rhetorical bridging can be seen in an interview with consumer behaviour analyst Joanna Feeley, in which she says: 'From car-sharing and bicycle-rental schemes, to sharing stories, habits and tastes across social media, people are increasingly inclined to share' (Roberts, 2012). Likewise, journalist Ben Bryant (2011) explains the growth of the sharing economy in terms of the success of social media: 'Not content with persuading us to share our photos, location and innermost thoughts with the world, entrepreneurs are now asking us to post our possessions on the internet and hire them out to neighbours.' Here, the online and offline are bridged by 'entrepreneurs', regardless of the fact that the entrepreneurs who are 'asking us to post our possessions on the internet' are by no means the same entrepreneurs who have been developing and promoting SNSs. Finally, an example from the *Washington Post*'s Katherine Boyle: she explains 'why it doesn't seem weird to share ties and toys with strangers' by means of the sarcastic comment that 'You've already shared your sonogram with 4,317 of your closest Facebook friends' (Boyle, 2012: 1). Once more, online sharing is presented as creating the conditions for offline sharing.[17]

The closest to an empirical assessment of this idea would appear to be a survey carried out by *Shareable* magazine together with the research company Latitude.[18] The survey claims to have found that 'Online sharing is a good predictor of offline sharing. Every study participant who shared information or media online also shared various things offline – making this group significantly more likely to share in the physical world than people who don't share digitally' (Gaskins, 2010: 2). Sharing information or media included music file sharing, use of Facebook and even sending a link for an interesting article to a friend. Sharing offline included using public transportation and sharing a living space, books or household items. Given this, one would be hard pressed to find any city-dweller under the age of thirty (or perhaps anyone who is not a hermit) who does not share. In any case, without the data themselves, it is hard to make clear sense of the claim that online sharing predicts offline sharing. What we can say is that the communicative and distributive senses of sharing are being treated as one, without acknowledging even the most basic fact that when I share media online, I do not have any less of the media, while when I share in the physical world my access to what I am sharing is limited.

This idea – that social media use can help account for the rise of the sharing economy – is also found in academic texts. In an important, and critical, piece on the sharing economy, Juliet Schor and Connor Fitzmaurice (2015) report that 'technophilia accounts for people's interest in connected consumption [Schor and Fitzmaurice's preferred term]. They like using the Internet to do things efficiently and easily' (p. 414). I have no argument with this. In addition, though, Schor and Fitzmaurice also claim that many people 'have been "digitally primed" by years of online sharing of files, music and other cultural products' (p. 414). I am unconvinced by this notion of 'digital priming' and think that the word 'sharing' is doing more work than it is given credit for. As with the Latitude report, here too one is left wondering why people who have been sharing files in a non-zero-sum game would therefore be keen on sharing material goods in a zero-sum game. It seems perfectly reasonable to suggest that people who have grown up interacting with other people through computers and mobile phones will feel comfortable using such devices to manage other aspects of their lives too, but why should it make them more interested in *sharing*, specifically? Put differently, the notion of 'digital priming' does not explain why the digitally primed should also be primed to act prosocially.

When they try to account for the connection that they posit between offline and online sharing, proponents of the sharing economy suggest two main ways that online sharing (of media and information; sharing on social media) drives or leads to offline sharing (of power drills and cars and spare bedrooms): first, the internet gives people a sense of community that they wish to reproduce offline; and second, the internet induces trust, which encourages offline sharing. Community and trust are doubtless important values for the sharing economy, as with sharing at large. For instance, one journalist posits that 'The services may be gaining popularity because they reinforce a sense of community' (Wortham, 2010: 15), and both Botsman and Rogers' and Gansky's books (Botsman and Rogers, 2010, esp. Chapter 4; Gansky, 2010, esp. Chapter 5) deal extensively with questions of trust. Regarding the purported sense of community, in an interview with London's *Evening Standard* newspaper, Roo Rogers makes the first of these points when he maintains that 'Social networks have reminded us of the importance of community', and that this helps explain the rise of the sharing economy (Bryant, 2011). J. David Goodman of *The New York Times* similarly writes that we are 'learning to share, thanks to the web', and notes that the internet is a 'communal' platform (Goodman, 2010: 2).

In terms of the production of trust, one way that social media are said to help is that they are 'a convenient credibility check (many sites require renters to log in through Facebook, as a guard against false identities)' (Baedeker, 2011: 31). Likewise, our 'online profile leaves a trail of digital bread crumbs that makes it harder to pull off a scam' (Wortham, 2010: 15). One commentator sees the sharing economy as 'a natural extension of social networking, which has created loose networks of people as well as, more importantly, trust in these networks' (Macken, 2011: 22). It would seem reasonable to suggest that people discover communities online (though of course many do not). However, the argument that, having experienced community online, people now wish to experience it offline (with different people), is not explicitly substantiated; it is speculation, with no real evidence behind it.

Another peculiarity in the causal linkage made by various commentators between use of online social media and participation in sharing economy ventures is that most of the latter involve interacting with an entirely different set of people from the former. If we think about peer-to-peer lending within a neighbourhood, for example, the assumption is that you do not know the person with whom you are interacting – and many would say this is precisely the point – whereas the majority of the people with whom we interact online, or at least in SNSs such as Facebook, are people whom we do know, albeit to varying degrees. In other words, if relations of trust develop between lender and lessee, they are *subsequent* to the sharing transaction, and do not precede it.

This is not to say that there is not a connection. Indeed, it would be obtuse to suggest that there is not. I am not going to linger over the obvious role of technology in the sharing economy, other than to briefly allude to the familiar insight from the science and technology studies (STS) literature that new technologies do not necessarily bring with them predefined uses (this was mentioned in the previous chapter). That is, just because smartphones enable the uses that support what we call the 'sharing economy' does not mean that the sharing economy is an inevitable consequence of smartphones.

Critiquing the 'Sharing Economy'

For many observers, one of the biggest problems with the sharing economy is its name, and I have already mentioned – more than once – that numerous commentators have insisted that 'it isn't really sharing'. What they mean by this is that we are using the wrong word. In *Verbal Hygiene*, linguist Deborah Cameron analyses such accusations (Cameron, 1995). We know, she says, that the relationship between words and what they signify is arbitrary (we are all Saussurians). In order for language to work, therefore, we agree to a 'linguistic contract' according to which we assent to the proposition that the 'true and proper purpose of words [...] is to correspond exactly to things in the "real world"' (Cameron, 1995: 150). When Uber and Airbnb are described as part of the sharing economy, therefore, people are exercised by what they see as the failure of the signifier ('sharing') to point to the correct signified. Despite repeated assertions that the sharing economy as we know it today is 'not really sharing', it could be that these complaints have come too late. As Robin Lakoff observes in relation to new words, 'Whoever gets to establish those connections [between form and meaning] first and best controls the meanings [... of...] new words' (R. T. Lakoff, 1990: 87).

I should stress that my concern here is not to insist that the term 'sharing economy' is necessarily the most accurate one to describe the range of interactions that it currently covers. Nor is it my objective to outright defend the application of the term 'sharing economy'; I am open to the possibility that some companies are engaging in 'sharewashing' (Kalamar, 2013).[19] What is particularly interesting, though, is to understand why people take umbrage at the term when they do (and why they do not take umbrage at it in other instances), and what this tells us about what they think sharing is and should be.

At a time when Uber had been coming under increased scrutiny for its labour relations and surge pricing policy, *New York Times* reporter Natasha Singer wrote the following:

> To be clear, I'm not objecting to the services themselves. Ride-hailing apps like Lyft and Uber, odd-jobs marketplaces like TaskRabbit, vacation rental sites like Airbnb, and grocery-shopping apps like Instacart have clearly made travel, lodging, home renovation and dining more efficient for millions of people. (Singer, 2015: BU3)

What Singer found 'problematic' was 'the terminology itself', and in particular the way that it 'frames technology-enabled transactions as if they were altruistic or community endeavors'. For this journalist, then, sharing involves altruism and community. Thus, her comments echo the popular conceptions of early childhood and humanity's prehistory that I described above.

What, then, of the suggestion that 'true sharing' (Belk, 2014) can be found in early human societies? A body of anthropological literature spanning over a century has examined practices of food sharing among hunter-gatherer societies. The classic position, according to Nurit Bird-David (2005), is stated by James Woodburn when he writes, 'Unquestionably sharing is of central importance in the operation of these societies' (Woodburn, 1998: 48, cited in Bird-David, 2005: 201). This would support those who say that the sharing economy 'is not really sharing': Look, they can say, *this* isn't sharing; *that* is.

This position, though, is questionable for two main reasons that can be extracted from the anthropological literature itself. First, as argued by Nurit Bird-David, our (bourgeois, western) concept of sharing does not readily map on to the practices of hunter-gatherer societies. Modern westerners see individuals as separate and discrete entities, making it hardly surprising that we should see game as property to be divided. However, the Nayaka – the hunter-gatherer society in southern India that Bird-David spent years studying – see it as 'a volitional sentient being' (Bird-David, 2005: 205). Also, although the meat from large game is distributed among the members of the Nayaka village, this is not about 'division into shares'. Rather, it is 'separation for joint consumption, and the manifestation and regeneration of relations' (p. 214). Furthermore, according to Bird-David, conceiving of the ways in which hunter-gatherers divide up food as sharing 'plays implicitly on bourgeois emotional chords, casting "hunter-gatherers" as a yearned-for "communitas"' (p. 204). This is very close to Benjamin's notion of an ur-past and, I argue, resonates with the way in which sharing – in its distributive moment – is imagined as prehistorical and natural. In other words, we impose our western and bourgeois associations of sharing – with property, labour and equality – on to hunter-gatherer societies, and then use their social practices, now defined by us as 'sharing', in order to define, in circular fashion, how we ourselves should behave. Moreover, this process takes place in the context of a modern yearning for an imagined and lost sense of community.

The second reason that hunter-gatherer societies should not be taken as inspirational instances of sharing communities is because, according to a large swathe of anthropological literature, they are not necessarily the instantiation of generosity and selflessness that our imagination would have us believe. If the common perception is that hunter-gatherers generously share food so that others will share later on, it is a perception of which we should be disabused (Woodburn, 1998). The force of this argument is to say that the same understanding of 'sharing' that, for some, makes the term 'sharing economy' a misnomer should lead us to question quite how oriented towards sharing the actual human societies that populate our ur-past were themselves.

Evidence is plentiful, and comes in two varieties. An anthropological classic of the first is Colin Turnbull's *The Mountain People* (Turnbull, 1972). In stark contrast to *The Forest People* (Turnbull, 1961), which tells of the BaMbuti (or Mbuti), pygmy forest-dwellers who live in harmony with one another, *The Mountain People* is about the Ik, who live in a state of distrust and discord. The Ik, says Turnbull, 'place the individual good above all else and almost demand that each get away with as much as he can without his fellows knowing' (Turnbull, 1972: 101). By way of illustration he relates how a tribesman to whom he had given some tobacco hid some of it before pretending to share it out among the others. In this vignette, sharing is a custom one must appear 'to publicly uphold' (p. 101), but not necessarily to practise. In another vignette, an Ik woman finds that half of the berries she had laid out to dry had been stolen. She accused her brothers, 'saying they were in league against her and this was what came of sharing food and trusting your relatives' (p. 174). These are not the sharing hunter-gatherers of yore.[20]

The Ik, then, serve as a counter-example to the prelapsarian natives that sharing economy proponents seem to have in mind. But even in societies that do appear to be sharing nicely, things are not necessarily as they seem. Returning to anthropologist James Woodburn, for instance, we learn that among the Hadza, a group that appears to share in what we view as a virtuous fashion, the division of resources is actually an obligation; the hunter does not control who receives meat; there is no stress on generosity and the donor is not thanked; receiving meat does not imply an obligation to reciprocate, and most people are actually unable to do so; and shares of resources are demanded. Moreover, Woodburn writes that 'If people can avoid requests to share, they will often do so. The most frequent way of doing this is by concealment of whatever it is that should be shared, and by telling lies or misleading other people about it' (Woodburn, 1998: 56).

Where does this leave us in relation to the assertion about today's sharing economy and the criticism of it that 'it is not really sharing'? One response might be to say that when appealing to 'age-old' practices of sharing, and evoking the timelessness of sharing (as instantiated by 'primitive' 'tribes' in some accounts), sharing economy proponents are making a kind of category mistake: their conception of sharing is too deeply embedded in our social, economic and cultural context for it to be meaningfully applied to societies so different from ours. Another response might be to say that the contemporary sharing economy is no less a sharing economy than that of the tribes whose practices are meant to be the prelapsarian standard for which we should be striving: they too are motivated by self-interest and have no true altruistic concern for others. (Of course, the riposte to this could be to say that hunter-gatherer societies are therefore not examples of sharing either.)

Another response might be to observe that sharing-as-distribution *always* takes place within some kind of socio-economic context. Indeed, as a social practice, that is unavoidable. The same is true of the nostalgic yearning for sharing (for an ur-past (Benjamin and Tiedemann, 1999), or for communitas (Bird-David, 2005)). 'Real' sharing is valorized as presocial: it is purportedly found among very young children and prehistoric humans (and the people our cultural imaginary holds to be their present-day representatives, remote hunter-gatherer tribes); it is said to be located within the family and among groups of intimates (Price, 1975); and for Belk (2007, 2010), alongside (bourgeois) family relations, mothering is a key prototype of sharing (after giving birth, the mother 'shares her mother's milk, nurturing, care, and love with the infant', he writes (Belk, 2010: 717)). However, setting aside what may be an over-romanticization of motherhood by Belk, even if breastfeeding ever was presocial, or a-cultural, today that is certainly not the case.

Similarly, Giana Eckhardt and Fleura Bardhi, authors of an important article that shows Zipcar users to be far from constituting a community (Bardhi and Eckhardt, 2012), make the case that 'The sharing economy isn't about sharing at all' (Eckhardt and Bardhi, 2015), on the grounds that 'Sharing is a form of social exchange that takes place among people known to each other, without any profit'. Sharing 'dominates particular aspects of our life, such as within the family', Eckhardt and Bardhi (2015) quite reasonably say. However, when they report that Zipcar users 'don't feel any of the reciprocal obligations that arise when sharing with one another', one wonders how that sits with the idea about the family as the prototype of sharing, which Belk, drawing on Price (1975), sees as fundamentally non-reciprocal (Belk, 2010).

One way of resolving this tension is by remembering that the family is not a zone that somehow lies outside market relations, as made clear decades ago by second-wave feminism. My point is not that mothers tot up a bill for breast milk as they feed their infants, or that children pay rent for their bedrooms – clearly the relations among members of a family are different from relations among work colleagues. Rather, our very view of family relations as 'natural' is one that should and can be contextualized (we probably need not look much further than Habermas (1989) to see this).[21]

This insight, together with the work of anthropologists who disabuse us of the notion of harmonious and sharing 'primitive' societies, implies that there is not a timeless, presocial version of sharing that we can meaningfully appeal to. This, though, is not to say that anything goes. It is to suggest that if we are to sustain a critique of the sharing economy – or those parts of it that look to us least like sharing – this critique should be developed internally to the sharing economy. This would exclude comparing sharing today with 'tribes', if only because hunter-gatherer societies live with a different relationship to money than we do. We could do this by considering the objectives of the sharing economy – reducing consumption and increasing a sense of community seem to be two widely agreed-upon objectives – and then asking whether particular instances embody that or not.

Conclusion

In this chapter, I have shown that the notion of a sharing economy first emerged in relation to production, more particularly peer-production, and even more particularly, digitally mediated peer-production. As we shall also see later in relation to file sharing, the digital origins of these practices made the application of the term 'sharing' to them fairly natural, though not necessarily neutral. By the time the sharing economy, and especially collaborative consumption, really took off in around 2010, SNSs had come to dominate the internet, expanding the notion of 'sharing' and enabling it to serve as a readily accessible metaphor. If online, social media sharing was digital and presented as reinforcing social ties (which many people still experience it as doing) and increasing a sense of community, then it was a short step to seeing the kinds of ventures that came to be known as the 'sharing economy' as sharing: they are digital, they are presented as reinforcing social ties and community, and they draw on already circulating distributive concepts of sharing. In other words, the metaphor of sharing in online contexts conveyed meaning from those contexts to the sharing economy, including conceptualizations of the internet as subversive, as well as conceptualizations of human interrelations based on trust, honesty and so on (the therapeutic sense of sharing, which I discuss at length in Chapter 6). This latter point is nicely instantiated by a request that participants at an event held by the pro-sharing economy group OuiShare not shake hands with one another, but hug.

Because all aspects of what we term the 'sharing economy' are linked metaphorically – through 'sharing' – to online sharing, and to sharing as a type of communication, we might expect to find commonalities in the practices that fall within its remit. And indeed we do, in that they represent an expansion of the public at the expense of the private. Another way of putting it might be to say that the sharing economy widens public access (sometimes for payment, sometimes not) to the private. This is how the 'sharing' of the sharing economy partakes in the contemporary metaphor of sharing; this is what is 'sharing' about it.

As a final thought for this chapter, I would like to suggest that saying 'it's not really sharing' might actually be a conservative position to adopt, at least among those who then do not go on to critique the economic model of companies such as Uber, or among those who define personal acquaintance and especially kinship as a condition for sharing. If the sharing economy is not really sharing because true sharing is what happens within families and intimate groups, then, by definition, there would seem to be no chance of taking sharing outside of those small social units. In other words, not only is the sharing economy apparently not a sharing economy, but no economy could truly be a sharing economy, apart from those interactions between family members in their home. According to this reading, sharing remains private; more than that, it remains feminine (Belk's example of motherhood reinforces this), and the modern affinity between womanhood, domesticity and nature is reproduced. Regardless of what today's sharing economy is or does, the notion of sharing held up by those who say it is not really sharing would actually seem to have the consequence of rendering *any* widespread sharing economy impossible. To me, at least, this seems like tacit support for the status quo.

5
Sharing Our Feelings

The concept of 'sharing' implies positive interpersonal relations. The way we create and sustain interpersonal relations nowadays is through talk. Sharing emotions, or conveying an inner state of ours to others, is crucial to what it means to be a proper, functioning member of society. Sharing, as a particular type of talk, can thus be conceived of as the constitutive activity of therapy culture (Füredi, 2004).

In this chapter, I explore the connection between digital culture and the therapeutic discourse. Perhaps in this chapter, more than any other, the reader will query my linking sharing as the dominant form of communication in therapy culture and sharing as the constitutive activity of social media (John, 2013a, and see Chapter 3). Are they really related? Does the fact that the word 'sharing' is prominent in both spheres really mean anything? And even if it does, so what? The objective of this chapter is to show that the therapeutic discourse (where 'sharing' is how we talk) and SNSs (where 'sharing' is how we participate) are related, and that the word 'sharing' helps us understand how. We have already seen, for instance, how SNSs make explicit use of therapeutic meanings of 'sharing' when they say things like 'Share with the people you care about'. However, the connection is deeper and touches on that which is considered private or public. Wuthnow (1994) has posited that '[s]upport groups are characterized by making private stories into public communicative acts' (p. 186). The ease with which we could replace 'support groups' with 'social media' suggests the commonality that this chapter seeks to explore. We shall be talking, then, about the collapse of boundaries on the one hand and the rise of emotional communication on the other. This is not to say that all communication on SNSs is of a therapeutic ilk, but it is to say that the most basic assumptions about personhood that we find in our therapeutic culture are also at play in that sphere, and that the notion of sharing links the two.

Social network sites are places where, *inter alia*, people express emotions. I take this to be an entirely uncontroversial statement. There are entire industries that strive to extract knowledge from the many expressions of emotion people make online (sentiment analysis, for instance), and academics too have been fascinated by a range of aspects of emotional communication online (see Benski and Fisher, 2013, for an excellent collection). The emotional impact of surfing has long been of interest too, from Kraut et al.'s (1998) findings (later moderated in Kraut et al., 2001) that correlated depression with internet use, to more recent research that suggests that Facebook use is correlated with social capital (Ellison, Steinfield and Lampe, 2007) and that emotional disclosure on Facebook can improve psychological well-being (see, for instance, Reinecke and Trepte, 2014).

For our purposes, however, we do not need to know much about who expresses emotions online, how, when and with what consequences. Of greater interest here is the notion that SNSs have an interest in their users expressing emotions on their platforms. As we saw in Chapter 3, 'sharing' is a mainstay of how SNSs market themselves. 'Sharing' incorporates the entire gamut of social media practices, and, through taglines such as 'Share your world' and 'Share your life', it is pretty clear that SNSs would like us to share everything. (Of course, as already discussed, sharing on SNSs is monetized and emotions are commodified; I shall come back to this aspect of sharing later on.)

Sometimes, SNSs appeal to us to participate by making explicit the association between sharing and emotions. Sharing on social media, they intimate, is good for us individually, and good for our interpersonal relationships. For instance, a post on Facebook's blog claimed that 'Facebook aims to reduce that very isolation Putnam [in *Bowling Alone* (Putnam, 2000)] laments by facilitating sharing with the people we care the most about', and unambiguously asserts that 'The more people use Facebook, the better they feel' (Burke, 2010). Recall from Chapter 3 the ad I mentioned for Yahoo!'s Pulse network (also from 2010), which encouraged us to 'Share what's important to you with the people you care about'.

My general point, and the theme of this chapter, is that SNSs are part of today's 'therapy culture' (Füredi, 2004); they partake in the therapeutic discourse (Illouz, 2007, 2008). I am not talking here about websites, forums and online support groups that are explicitly dedicated to their members' psychological welfare, but rather about the assumptions regarding the self and the self's relations with others that underlie social media.

To illustrate briefly what I mean, before engaging with the concepts of therapy culture and the therapeutic discourse more intensively, consider a video produced by Facebook and published on Facebook's own Facebook page on 12 February 2015 (two days before Valentine's Day).[1] The text that accompanies the video reads 'Love is for sharing'. One meaning of this is that if you have love, you should share it around; in this sense, 'Love is for sharing' is a play on the more familiar phrase 'Share the love'. But of course what Facebook intends here is that we should share our love by communicating about it on Facebook, and this works as a tagline because 'sharing' is the word we use to refer both to the communication of emotions and to the act of posting to SNSs. In this short clip (1:06 minutes), we see a number of characters talking about love, directly or indirectly, and reading out status updates (including saying the words 'heart emoji'). Finally, the message of the clip is that if you love someone, tell them (on Facebook). The central example in the clip is of a mother's love for her young adult daughter and the daughter's love for her mother, though other instances of love appear, including (hetero- and homosexual) romantic love. However, relations between the self and other are not the only focus of the clip, as one of the talking heads declaims the famous Oscar Wilde quotation, 'To love oneself is the beginning of a life-long romance.' In other words, we should have high levels of self-esteem (a particularly important concept in what Moskowitz (2001) calls the 'therapeutic gospel'). In this clip, then, we are told that we should communicate our emotions to significant others, and that our own self and happiness should be at the forefront of our concerns.

These are two important features of what it means to live in a therapy culture. We shall now turn to the first of these, namely, the injunction to share our emotions.

Sharing and the Therapeutic Discourse

As a first step, let us observe the word 'sharing' in its sense of talking about emotions. In her analysis of the culture of therapy and self-help, sociologist of emotions Eva Illouz (2008) does not analyse the word 'sharing' per se, but the examples she cites show how it is used. Thus, she quotes a source as explaining, 'Sex is a very intimate encounter, one that involves sharing feelings' (p. 127). Illouz herself says that 'the therapeutic ideal increases the injunction to share all needs and feelings' (p. 227). In another important book about the rise of the therapeutic discourse, Deborah Lupton lays out a cultural requirement of our intimate ties: 'Individuals in close relationships are expected to achieve and maintain intimacy by sharing their emotions with each other, even if these are negative' (Lupton, 1998: 96). Bellah et al., in their classic, *Habits of the Heart*, also use the word 'sharing' to describe the idea of love in a therapeutic culture: 'Thus sharing of feelings between similar, authentic, expressive selves [...] becomes the basis for the therapeutic idea of love' (Bellah, Madsen, Sullivan, Swidler and Tipton, 1985: 100).

As competent social actors, we know that this is an important meaning of 'sharing', and we most likely adhere to the injunction to share in this sense in our own relationships. But, as Chapter 2 showed, there is a history here. In the introduction to *Therapy Culture*, Frank Füredi points to the emergence of a therapeutic system by showing how terms such as 'stress', 'trauma' and 'self-esteem' have become far more prevalent in the press (Füredi, 2004: 3–7). Similarly, put the term 'share your feelings with' into Google's Ngram interface,[2] and you are presented with a cultural shift (see Figure 5.1). During the second half of the 1960s, the term takes off. This sits well with what we know about the 1960s and 1970s – the counterculture, the rise of therapy and support groups, consciousness raising of various kinds, and the growing popularity of TV talk shows. The countercultural radicals of the 1960s 'wanted to organize all human relationships along therapeutic lines' (Moskowitz, 2001: 179) and 'liberate the self' (p. 217). In the 1970s, self-expression was seen by many Americans as the key to liberation, with 'candor, intimacy, and self-awareness' especially lionized (p. 219). According to this generation, 'Salvation', argues Moskowitz, 'lay in openness and communication' (p. 219). In Moskowitz's historical account of therapy culture, the 1970s saw an unprecedented focus on self-expression and feelings. 'Never before had such an emphasis been placed on intimacy, trust, and self-awareness' (p. 243). If 'sharing' conjures up rainbows, this is when that link was made.

This is an important part of what Paddy Scannell has termed the 'communicative turn' of the postwar western world (Scannell, 2009). It represents the consolidation of a self that is constructed through talking or writing about itself, be that in therapy, support groups, on television talk shows or in everyday interactions with significant others and even colleagues. It is a self that is (relatively) free from social shackles, at least formally so (parents were not choosing their children's marriage partners in 1960s America, for instance). This self, then, is free to engage in 'pure relationships' (Giddens, 1992) that are based on the verbalization of feelings; or, as Beck and Beck-Gersheim put it, 'What used to be a team sharing the work [i.e., pre-modern agricultural families] has turned into a couple sharing emotions' (Beck and Beck-Gersheim, 1995: 48).

Beck and Beck-Gersheim, like Habermas (1989) and Giddens (1992), trace this shift back to the beginning of modernity, but it would seem that the turn of the twentieth century was a key period in the internalization of the value of talk (see T. J. J. Lears, 1983; Moskowitz, 2001). In particular, I want to shine a spotlight on a group of religious Americans whose spiritual practices revolved around 'sharing', namely, the Oxford Group, which, given its status as the

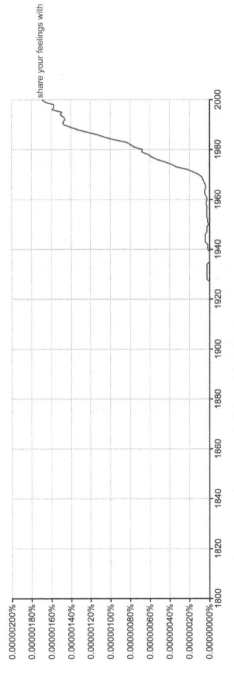

Figure 5.1 Google Ngram, 'share your feelings with', 1800–2000

progenitor of Alcoholics Anonymous, is surprisingly unfamiliar to scholars of communication and therapeutic culture. As we shall see, the Oxford Group's use of the word 'sharing' was taken up by Alcoholics Anonymous and is now used extremely widely. In the first social scientific treatment of the Oxford Group, Allan Eister wrote:

> New religious movements are a source of perennial interest to the social scientist. They are the stuff out of which new institutions are sometimes formed. Often they open up fresh cleavages in society (or uncover latent ones); and not infrequently they have added important new values and patterns to the culture. At the same time these new movements bear the marks of the social order out of which they spring and are themselves products of significant social trends operative in society. (Eister, 1950: ix)

There is no doubt that the Oxford Group gave birth to new institutions (Alcoholics Anonymous), and I hope to show that it 'added important new values and patterns' to western culture – at the very least, the appellation of a certain type of talk, 'sharing'. I shall also attempt to place the Oxford Group in some kind of context, though with the advantage over Eister of an extra half-century of hindsight.

Sharing and the Oxford Group

The Oxford Group was founded in 1922 by Frank Buchman. It was a Christian group that, according to Dick B., a historian of Alcoholics Anonymous, aspired 'to get back to the beliefs and methods of the Apostles' (B., 1997: 52). The Oxford Group (no relation to the university) was the forerunner of the Moral Rearmament movement that was popular in the US in the lead-up to and during the Second World War. More significantly for our concerns here, it was also the progenitor of Alcoholics Anonymous, and thus in a way the forefather of today's wealth of support groups. This is important because, as Dick B. wrote, 'If there was one Oxford Group practice that had more influence on *today's* A.A. meetings and message-carrying ideas than any other, it was the practice of *sharing*' (p. 64; emphasis in original). The place of the Oxford Group in the history of AA is also emphasized by Bill Pittman in his history of the organization, where he writes, 'There is little doubt about the Oxford Group's contributions in influencing the formation of Alcoholics Anonymous' (Pittman, 1988: 174). To support this, he cites AA co-founder Bill Wilson's own history, in which he writes, 'The early AA got its ideas of self-examination, acknowledgment of character defects, restitution for harm done, and working with others straight from the Oxford Group' (W., 1979: 39; cited in Pittman, 1988: 175).

As mentioned, a central practice of the Oxford Group was sharing. In a pamphlet, *What is the Oxford Group?*, penned by the anonymous 'Layman with a Notebook', we read that 'The Sharing of our sins and temptations with another Christian life given to God' is the first of the Group's four spiritual activities which, it is claimed, will help people live according to the principles of 'Absolute Honesty, Purity, Unselfishness, and Love' (The Layman with a Notebook, 1933: 8). 'Sharing of sins as practised by the Oxford Group', according to the Layman, 'is sharing in the ordinary sense of the word; in plain language it is telling, or talking over, our sins with another whose life has already been surrendered to God.' As we saw in Chapter 2, 'sharing' already meant telling by the 1920s – this is the 'ordinary sense of the word'. But the practice it referred to was new, and the spread of the practice, along with the word, gave the word a new set of meanings.

While sharing in the Oxford Group also included listening to another's confession, the dominant sense was that of confessing. Unlike the Catholic confessional, however, sharing in the Oxford Group was a public act carried out in a secular setting (someone's living room, for instance) in front of a group of laymen (as opposed to privately, in a confession box, to a priest or even more privately to God). For the first time, then, 'sharing' meant the public profession of a fault of self (sin) for the sake of redemption. Indeed, the very origins of the movement, according to its founder, Frank Buchman, had their foundation in a religious experience he had where he felt that his heart obtained release from bitterness through the act of telling other people how he had maltreated them

(B., 1997). Talking openly and honestly to others, we might say, increased Buchman's subjective well-being.

What is the Oxford Group? (The Layman with a Note-book, 1933) and *For Sinners Only* (Russell, 1932) are both very useful texts from which we can extract a great deal of the values of the group, and in particular the implicit theory of communication and well-being that underpinned its emphasis on sharing. For this reason, I shall linger on them and present a number of extracts which, I believe, demon-strate the group's psychological bent (despite its strident opposition to psychotherapy) and show how its concept of sharing was a profoundly therapeutic one.

Given its stress on interpersonal relations, the Oxford Group was careful not to undercut or devalue the individual's unmediated relationship with God, but it is still hard not to read it as a modern organization with secular undertones. As the Layman wrote, it is enough that we acknowledge 'a need for change in our lives' in order to get help from the 'life changers' of the Oxford Group (p. 5); here, I would argue that the very ability to think in terms of instigating a 'change in our lives' is post-traditional. In this same pam-phlet we also read what would appear to be a criticism of positive thinking, part of the group's broader rejection of psychology, at the same time as it irrefutably adopts a psy-chological model of the self and interiority. Regarding our sins, the Layman writes that, 'We cannot get rid of them by deciding to think no more about them' (p. 21). Rather, they must be 'cut out by a decided surgical spiritual operation', otherwise 'we become warped in outlook not only towards others but towards ourselves' (p. 21). Later, in the chapter titled 'Sharing', he talks about the value in 'getting things off our chest' (p. 28). Sharing (within the context of the group) is described as better than psychotherapy (with which it is clearly in competition):

> A father who Shares with his son the knowledge that he, too, went through the phase that his son now finds so torturous to his physical and spiritual outlook: an elder brother Sharing with a younger, a friend with a friend, can lay simply and easily those ghosts which the psychoanalyst can only eradicate by prolonged effort. (1933: 31)

But why not confess to God? What are the properties of sharing with other people that ensure redemption? First, notes the Layman, people can of course 'go straight to God' and 'God will forgive them' (p. 32). However,

> if they wish for a sure and certain knowledge that their past sins – and all of them – are to be wiped out, once and for all these sins must be *brought into the open* and honestly faced. *To put them into words*, before Christ with another Christian, as a witness, is the only healthy way of making sure that the spiritual system is virtually cleansed. (Ibid.: 32; emphasis added)

The italicized phrases here are crucial. Work on the self cannot be conducted internally. Our sins (the modern reader might substitute 'faults', 'neuroses', etc.) can only be redeemed (resolved, cured, worked through and so on) by being made public or, at the very least, by being brought outside of us, exteriorized, so that they can be observed by another. This interpretation is supported by an extract from *For Sinners Only*, in which Russell (1932) discusses the place of God in sharing. Ideally, he says, confessions 'should be made direct to God'. However, 'there are very many who need the help of Sharing with another [...]. For them Sharing is a practical necessity. Only so do they grasp the *reality* of their confession' (pp. 16–17; emphasis added). This notion is reinforced by a quotation from a conversation between Russell and Canon Prof. Grensted, a scholar and psychologist:

> 'Sharing is a positive good,' emphasised the Professor. 'It is the real answer to so many problems. For *it is almost impossible for a person to see his real problems straight unless he has an outside view on them*, and he can only get that by Sharing.' (Russell 1932: 238; emphasis added)

Eister (1950), the social scientist, presents a lengthy excerpt from Philip M. Brown (1937), an academic and Oxford Group member, on the subject of sharing and people's relationship with God:

> I have found that to deal drastically with sins it is necessary to share them completely with someone in whom we have

confidence. *Few persons are honest and definite with God.* There is an inevitable tendency to be vague when we confess our sins alone. We do not name them in ugly detail. We do not lay them out as it were on a table and *look at them objectively.* [...] We cannot possibly see our sins in their true light [...] until we 'share' deeply with another human being (Brown, 1937: 33–4; cited in Eister, 1950: 18–19; emphasis added).

Once again we are presented with the idea that truth is ensured not by talking to God, but by placing ourselves under the scrutiny of another person. This presages the logic both of talks shows and reality TV, which also have a deeply therapeutic bent, and of social media.

Truth is attained via the gaze, and putting our 'sins' into words both objectifies and reifies them, making them real. This enables others to gaze upon our sins and confessions (or upon us) and thus redeem us, or give us value. This is what Illouz is referring to when describing a much more recent technique for partners to let off steam: 'This technique', she writes, 'instructs that we contain negative emotions and make them into emotions external to the self, to be watched from the outside, so to speak' (Illouz, 2007: 35). We shall come back to the idea that truth is attained through surveillance when we touch upon the place of reality TV in contemporary therapeutic performances of the self. For now, we note that sharing for the Oxford Group entails lowering the boundaries around the self and, as cited above in the context of support groups, 'making private stories into public communicative acts' (Wuthnow, 1994: 186).

Another aspect of sharing, as emphasized by Russell and others in the Group, was that of fellowship. Sharing, said Russell (1932), was 'the only way to true fellowship' (p. 17). Moreover, the founding story of how sharing was developed in the Oxford Group was one of the formation of fellowship: Buchman and some other men were travelling across China by train, preaching. Living in close quarters, tensions sometimes rose, which were resolved by sharing annoyances 'with the person concerned rather than with another'. The men 'share[d] our own faults' which enabled them to 'cement the team into the ideal fellowship' (p. 196).

Perhaps unsurprisingly, sharing is also held to be the bedrock of marriage. In another contemporary text about the Oxford Group, Victor Kitchen (1934) says that it is the 'sharing of one's sins' that turns 'an ordinary marriage' into 'holy wedlock' (p. 61). Russell (1932) tells of the tribulations of a married couple for whom sharing resolved a profound conflict. The wife had trained as a lawyer before marrying and having twelve children; now, however, she was unhappy. After a long conversation with an Oxford Group leader, she realized that 'she must just share the whole story with her husband'. The husband was deeply shocked to hear of his wife's unhappiness. However, 'It ended in both offering their lives to God, and they are rebuilding their home on under-standing lines of simply Sharing. The wife has become a lawyer, and is practicing law with the whole-hearted assis-tance of her husband' (pp. 232–3). Russell then cites a man who explains his view of marriage:

> This team-work for Christ demands an open transparency between husband and wife, with no hidden reservations about each other. Most people are lone wolves by nature, and they need a regular time of Sharing each day, when all the silly little irritations, the jealousies, the suspicions which skim the cream from matrimony can be laughed away. (1932: 234)

Most of us today would consider this sound advice, and indeed very similar modes of talk are reported fifty years later by interviewees in *Habits of the Heart* (Bellah et al., 1985), pointing to a continuity both in terms of language and cultural values. Many couples, we learn in *Habits of the Heart*, 'speak of sharing – thoughts, feelings, tasks, values, or life goals – as the greatest virtue in a relationship'. An inter-viewee talks about how she fell in love with her husband – '"I think it was the sharing, the real sharing of feelings"' – and the authors conclude that in American culture, the 'natural sharing of one's real self is, then, the essence of love', no less (p. 91). Observing patterns of communication among couples at around the same time as Bellah et al. were writing, Mary Anne Fitzpatrick noted that 'In the latter part of the 20th century, the sharing of personal feelings and information has become the hallmark of a close relationship' (Fitzpatrick, 1987: 131).

Notably, sharing in the Oxford Group was criticized in ways that resonate with contemporary critiques of the therapeutic ethos and sharing as its primary form of communication. In a book based on fifty-five case studies of current and former Oxford Group members, Walter Clark (1951) lays out some of the frequently expressed criticisms of the Group. One of these is that 'it is blind to social evil, and by its complacency is helping to perpetuate the shortcomings of the present social order' (p. 233). Likewise, in Eister's early sociological account, the Group was described as focusing on 'the moral shortcomings of individuals with little or no attention focused upon the institutional mechanisms and processes operative in the social order' (Eister, 1950: 184). This reads very much like a prefiguration of later critiques of the therapeutic ethos, such as those of Rieff (1966) and Lasch (1978), who bemoan the decline of the social at the expense of the individual.

Relatedly, Clark's interviewees who viewed the Oxford Group in a negative light saw the Oxford Group experience as overly emotional and not sufficiently intellectual (Clark, 1951). An interviewee whom Clark categorized as 'partly positive' nonetheless criticized the Group's 'over-emphasis of the emotional side of religion, in the unrestrained public confession of sins' (p. 167), and even in the chapter comprised of positive experiences with the Group one interviewee referred to 'its too great dependence on emotion' (p. 178).

There are even indications that 'sharing' was a manipulative tool. Clark (1951), for instance, writes that a 'member of the Group found a premium coming due on his insurance policy, his only monetary asset. He was "guided" [in Quiet Time, by God] to share his financial worries with another wealthier member, with the result that the premiums for the next five years were paid for him' (p. 30).

While I would not posit that today's social media sense of sharing is the same as the Oxford Group's, I link the two through the model of an individual self who conveys his or her authentic being through words that are placed in the public domain. What I show is that the kind of communication that SNSs encourage us to engage in has cultural roots in the kind of communication that the Oxford Group called 'sharing', which in turn draws on an understanding of the self that was emerging in the early twentieth century.

'Sharing' as Talking about Emotions

It is not easy – perhaps it is impossible – to definitively establish that our current understanding of 'sharing' as talking about our emotions is based in the Oxford Group, though this suggestion would seem to be supported by the corpus analysis presented in Chapter 2, as well as by a lack of references in texts about the Oxford Group to other people meaning something similar by 'sharing'.[3] It is far easier to establish that sharing is a form of communication that is central to how we live today, and especially to our interpersonal relationships. I shall attempt to do this by drawing on three key texts: Katriel and Philipsen's (1981) article about communication as a type of talk; Carbaugh's (1988) book about the talk show *Donahue*; and Cameron's (2000) critical appraisal of communication.

What we today call 'sharing' used to be called 'communication', as distinct from 'mere talk' (Katriel and Philipsen, 1981). For Katriel and Philipsen, ' "communication" refers to that speech which manifests self-disclosure, positive regard for the unique selves of the participants, and openness to emergent, negotiated definitions of self and other' (p. 315). Katriel and Philipsen distinguish communication from 'mere talk' by characterizing the former as close (as opposed to distant), supportive (as opposed to neutral) and flexible (as opposed to rigid). Katriel and Philipsen state that closeness between interlocutors is manifested 'To the degree that each interlocutor makes public what was previously private information about his or her unique self image' (p. 308). Second, 'communication' is 'speech in which unconditional positive regard finds its natural home' (p. 308). This does not mean that we approve of everything our interlocutor has done, but that we 'approve the other *qua* unique and precious individual' (p. 308). There is something fundamentally non-judgemental at play in this type of speech. Third, 'communication' is characterized by flexibility; it is 'the speech of emergent realities, of negotiated selves and the negotiated relationship' (p. 308). Today, I would maintain, these features apply to 'sharing', which has largely supplanted 'communicating' as the type of speech we are engaged in when talking about our emotions and selves. The similarities with the Oxford Group's 'sharing' are also fairly plain to see.

In characterizing 'communication', Katriel and Philipsen refer to the television talk show *Donahue*, which is also the subject of Donal Carbaugh's book *Talking American*, which contains the only – to the best of my knowledge – treatment of sharing as a category of speech (Carbaugh, 1988; see esp. Chapter 8). Like Katriel and Philipsen before him, Carbaugh also stresses closeness and support as constitutive of this cultural category of speaking. Sharing, says Carbaugh, is 'making resources of self available to others in a tone that could support and benefit them.' It is 'making something that is one's own, for example, one's "psychological state," available to others' (p. 144).[4] Sharing, he says, 'is not only an expression of one's inner experiences and feelings, but is also speech with a relational embrace' (p. 145). Referring to men in a support group, Carbaugh says that 'sharing' is a type of talk through which 'The men gain support from one another'. They are sharing when 'they speak in a cooperative way about their common problems' (p. 147). Thus, argues Carbaugh, 'the primary purpose of sharing is expressive and affiliative' (p. 148), echoing the closeness and support of 'communication' in Katriel and Philipsen. Katriel and Philipsen's 'flexibility' is also present in that the talk show is held to be emotionally transformative, for both the guests and the audience.

In another echo of Katriel and Philipsen, and to recall once again the cartoon of the man 'sharing' beer with a child, Carbaugh notes that 'Socially discreditable behavior is overlooked and ignored when speech is labeled sharing' (p. 148). Here Carbaugh adds another aspect to this discussion: the fact that the way the speech is labelled impacts on how we view it. Calling talk 'sharing' is to use the word as an 'illocutionary force indicating device' (Searle, 1969), or as a 'metalinguistic performative' (J. Thomas, 1995). In other words, if we talk about a type of talk as 'sharing', we are creating an expectation for personal, emotional and intimate talk; if someone tells us they have something they want to share with us, we will expect to be told something of personal import, and not that the toothpaste has run out.

Building on Katriel and Philipsen and Carbaugh's analyses, Deborah Cameron adds the notion of 'therapeutic discourse'. Here, the key idea is that talking is *the* way to solve personal and interpersonal problems. Cameron herself does not call this talk 'sharing' directly; rather, she encloses the word in quote marks, presumably because she adopts an explicitly critical standpoint towards this type of talk (Predelli, 2003), and, as we know by now, if we call something 'sharing' we are saying it is good. However, it is clear that she is talking about the same kind of talk that so interested Katriel and Philipsen and Carbaugh before her. Cameron notes that one of the rules of the therapeutic way of talking is that one talk 'as an autonomous individual subject whose experiences and feelings are unique' (Cameron, 2000: 156; this is what American preschoolers learn through participation in the 'sharing circles' mentioned in the Introduction). Moreover, therapeutic talk is talk that can help one change for the better, recalling Katriel and Philipsen's concept of 'personal growth' (Cameron, 2000: 158), and similarly to the Oxford Group's hopes for people who shared their sins. Moreover, knowing how to talk this way is crucial for the ability to maintain intimate interpersonal relationships. Referring back to Carbaugh, Cameron also discusses the expectation that one will not be judged by what one says when therapy-talking.

Before examining how this therapeutic sense of sharing is played out in online environments, let us briefly review the common threads that connect sharing in the Oxford Group with sharing as talking about feelings today, so that they may inform our discussion of online sharing-as-telling.

First, sharing is authentic and reveals one's innermost self by conveying emotional rather than intellectual content – relatedly, therapists are more interested in how we feel than in what we think. Through sharing in the Oxford Group, we are told, 'People come to know one another as they really are' (Clark, 1951: 35). Members of the Oxford Group were meant 'to share our real selves, *sins*, and all, with each other' (Kitchen, 1934: 96; emphasis in original). In a Goffmanesque description of a meeting at Oxford University, Group member and author A. J. Russell wrote: 'Young men were revealing their real selves, [...] were showing a masked world how to be honest by removing their own masks' (Russell, 1932: 68).

This feeling of absolute honesty was also described by Philip M. Brown:

> This is the heart of sharing: one human soul going out to another in all humility, generous understanding, and confident faith, so completely and fearlessly that for once in his life a man can know the immense relief of being absolutely honest without reservation or concealment. (Brown, 1937; cited in Eister, 1950: 19)

Relatedly, sharing in the Oxford Group was very individualistic, and 'had almost exclusively as [its] focus the individual and some inwardly felt shortcoming' (Eister, 1950: 171–2).

Second, such sharing contributes to strong interpersonal ties, which in terms of the Oxford Group is called 'fellowship'. Indeed, the very circumstances in which the practice of sharing was developed involved close interpersonal relations. As mentioned, the story goes that Buchman and some companions were travelling in close proximity through China, and that being open and honest about how they felt towards each another not only prevented the group of missionaries from splitting up, but actually brought them closer together. Oxford Group member Victor Kitchen testified to the sense of friendship that sharing created between him and others on a retreat by noting that 'after but three days of sharing and fellowship in Christ I grew to know and love each one of the forty [people on the retreat] as I had not known or loved my own chosen clubmates during an intimacy of years' (Kitchen, 1934: 78). According to Clark, this fellowship comes about precisely because 'People come to know one another as they really are' (Clark, 1951: 35). Sharing is a profoundly honest type of communication.

Third, when listening to someone sharing, one remains non-judgemental. In the Oxford Group one would share in order to gain redemption and whoever was listening helped in that, simply by listening and accepting, or perhaps even by identifying with the sins being confessed and adding their confession too. As the Layman (1933) put it, at Oxford Group house-parties discussion took place 'directly and constructively without partisanship' (p. 11). Similarly, Russell (1932) explained that 'Changed men might go wrong in trying to change others by argument, but they were on safe ground in *recounting their own experiences* as the Apostles recounted theirs' (p. 22; emphasis added). This sense of a lack of judgementalism, or a feeling that sharing is a type of talk that does not invite argument or contention, is concisely conveyed by the response in Alcoholics Anonymous after a group member has finished talking, namely, 'Thank you for sharing.' Carbaugh (1988) deals with the potential pitfalls of this non-judgementalism when discussing the racism expressed by a guest on *Donahue*. Donahue says of her, 'At least she is being honest. [...] She is sharing her feelings' (p. 148), as if this can compensate for the utterance of a racist remark. Deborah Cameron (2000) confronts this sense of a lack of moral grounding more explicitly. In discussing the type of speech on shows such as *Donahue*, she says that it is talk that is 'governed by a norm of "righteous tolerance" according to which everyone has a right to their opinion, but it is not legitimate to impose your opinions on others or to judge them' (p. 153). Sharing, then, is expressing one's personal truth in a way that cannot be argued with.

Fourth, there is a distinct sense that when Oxford Group sharing is described, the text could be referring to therapy as we know it today. For instance, A. J. Russell relates a conversation he had with another Group member over dinner: ' "What is the Group teaching about smokes and drinks?" I asked. "What do you think?" That is the characteristic Group answer. [...] Throw a question at the Group and it comes back to you' (1932: 65–6). I have already mentioned the value noted by the Layman (1933) in 'getting things off our chest' (p. 28). He also says that talking about our sins 'brings them into the open light', meaning that 'they are no longer bogies but hard facts to be faced squarely' (p. 29), and stresses that sharing is not about 'Placing the blame on others and making excuses for our weaknesses' (p. 33). Moreover, the excerpts about marital harmony, and the story of Buchman and his companions learning to get along during their journey in China, sound as if they were lifted from present-day manuals for how to successfully maintain relationships. Members of the Oxford Group themselves were not unaware of this. Philip M. Brown, for instance, said quite unequivocally that 'The method of sharing or of "confession" is good psychology' (Brown, 1937; cited by Eister, 1950: 19).

Finally, and by way of extending the previous point, we can see similar views about the price of not sharing in both Oxford Group texts and contemporary common sense, and specifically the notion that a failure to share has profound negative psychic consequences. Russell (1932) again:

> What mattered far more was the danger of not Sharing. The life shut in on itself fell into disaster. The danger of Sharing could never be so great as the danger of bottling up. This led to all kinds of tragedies – including suicide and murder. (p. 238)

Jumping forward forty-five years, Susan Sontag (1978) described the power of such reasoning in folk attitudes towards cancer: 'many people believe that cancer is a disease of insufficient passion, afflicting those who are sexually repressed, inhibited, unspontaneous, incapable of express-ing anger. [...] cancer is now imagined to be the wages of repression' (p. 21); it is viewed as a disease of the 'failure of expressiveness' (p. 48) and related to 'emotional withdrawal' (p. 55). While views such as these are probably far less preva-lent than they were in the 1970s, I would wager that there is not a reader of these lines who would not encourage a friend who was feeling down, or who had gone through a difficult experience, to talk about it, perfectly convinced that things will only get worse if they do not.[5]

Mediated Displays of Authenticity

The point of devoting this much time and space to the com-municative aspect of sharing – sharing as talking, specifically about our inner selves as part of intimate relations – is twofold. First, it enables us to see the history of the concept in this particular context, a history that has not yet been told. Second, it highlights that talk is always tied to a cultural context, and that cultural contexts always have a political economy.

In her book *In Therapy We Trust*, Eva Moskowitz (2001) notes that the advice column in *Good Housekeeping* in the

1890s was slow to take off because what she calls the 'therapeutic gospel' had not yet been widely accepted. People feared self-revelation and did not want to expose themselves. She also discusses the psychologization of poverty and the new theory of welfare in the 1920s and 1930s that were behind reforms to mental hospitals in the US. However, a new treatment, 'quaintly called "talk"' (p. 43), took time to be adopted, as not all patients were 'accustomed to the idea, readily accepted today, that they should disclose intimate personal facts to perfect strangers' (p. 43). This idea – that talking about yourself provided access to your core being and could improve your well-being – took hold of the general public at a time when authenticity itself was being questioned and hence reconstructed.

This is the thrust of T. J. Jackson Lears' (1983) seminal work on the therapeutic roots of consumer culture, which he locates in the years 1880–1930 – the decades leading up to the flourishing of the Oxford Group. I shall not here be able to do justice to Lears' analysis, but rather point to some key points. To start, urbanization and secularization, argues Lears, contributed to a disturbance in people's sense of the real. As life became more comfortable (through technological advances such as canned food, central heating and plumbing), it came to feel 'thinner'. At the same time, the city produced a sense of anonymity – this has been broadly documented by early sociologists – which saw personality trump character, according to Lears. Similarly, the relatively new social requirement that people tune and adapt their behaviour to the unfamiliar others surrounding them produced an empty sense of self; all that was left was impression management. Meanwhile, processes of secularization saw 'ethical convictions [grow] more supple', while 'experience lost gravity' (p. 10). Significantly for the later success of the Oxford Group, even liberal Protestant leaders were urging that people must look inward and spoke of the sanctity of human potential. In short, writes Lears, the emergent therapeutic ethos was related to 'personal quests for selfhood in an ambiguous moral universe' (p. 29). The decentring of the self produced a drive to resolidify it. Lears takes these to be the cultural conditions that made possible a consumer culture that appealed to new conceptions of the self.

Of particular interest here, though, are the new 'longings for reintegrated selfhood' (Lears, 1983: 17) and the 'quest for self-realization' (p. 29), and in particular the way these longings have provided the basis for critical analyses of reality TV, which, in its therapeutic moment, has significant commonalities with social media and SNSs. Although much has been said about the instantiation of the therapeutic ethos in talk shows (Carbaugh, 1988; Illouz, 2003; Livingstone and Lunt, 1994, are just three prominent examples), I wish to proceed directly to reality TV, for which '[t]he question of emotion-based authenticity is crucial' (Aslama and Pantti, 2006: 177).

Bridget Griffen-Foley (2004) has pointed to 'considerable historical resonances' (p. 544) between reality TV and earlier forms of audience participation (advice columns in *Good Housekeeping* have already been mentioned in this regard). Similarly, as Beverley Skeggs has noted, ' "Reality" television is located within the longer process of modernity whereby personhood is opened out through the display of intimacy' (2009: 73). Her use of quote marks around the word 'reality' reflects the question that everyone – viewers and researchers alike – asks themselves about reality TV: how 'real' is it? Everyone knows there is contrivance (indeed, Rose and Wood, 2005, report that its 'contrived authenticity' (p. 294) is precisely one of reality TV's most attractive characteristics), which raises the question, what do reality TV shows stake their claim to reality on?

For Andrejevic (2004), it is the constant surveillance of reality TV participants (especially in shows such as *Big Brother*) that guarantees their authenticity. No one, when subjected to that much surveillance, could keep up a facade, we think, and so the surveillance itself promises truth, reality and authenticity. The expression of emotion, especially negative emotion, preferably accompanied by tears, is another guarantor of authenticity. Accordingly, for Aslama and Pantti (2006), 'the confessional monologue is one of the genre's main features' (p. 168), and has become 'the stage for emotional expression and self-disclosure' (p. 172). This is very much the premise of a study of reality TV shows about tattooists and their clients (Woodstock, 2014). Central to Woodstock's analysis of shows such as *Miami Ink* are the stories that tattooees tell about their tattoos and why they wanted that particular image. When telling those stories, writes Woodstock, they are narrating themselves 'verbally in a therapeutic language that focuses on individual suffering and healing' (p. 782). Moreover, precisely because of the centrality of the display of emotions to reality TV, participation in a reality TV show can be an experience that contributes to one's self-understanding. As Andrejevic says, in reality TV, '[V]oluntary submission to comprehensive surveillance becomes a therapeutic experience' (Andrejevic, 2004: 86).

I think it is fairly uncontroversial to state that, whatever else it may be, our participation in the platforms of social media is also an example of our '[v]oluntary submission to comprehensive surveillance'. Given this, a number of scholars have explored the similarities between reality TV and social media, both of which are examples of what Jon Dovey (2000) has called 'first person media'. Stefanone, Lackaff and Rosen (2010), for instance, posit that 'SNSs provide a unique platform for the reproduction of behavior observed in and modeled by [reality TV] programing' (p. 513). They see reality TV, in which 'actors regularly engage in "confessions" where they ritualistically disclose their private thoughts and feelings to the broadcast audience' as similar to online communication platforms, which have 'likewise enabled a growing number of Internet users to publish their thoughts, photos, and videos on the Web' (p. 509). Similarly, Rachel Dubrofsky (2011) draws on her own previous work on reality TV (Dubrofsky, 2007) in suggesting that surveillance is the key common ground between reality TV and Facebook, both of which, she argues, 'are forms of media that gained immense popularity in the last decade, creating spaces where subjects are constructed through the mediation of technology that does the work of surveillance' (p. 113). Showing how productive a focus on disconnectivity can be, Laura Portwood-Stacer (2013) reports people accounting for their refusal to use Facebook by reference to reality TV: she cites a website in which a university professor equates sites such as Facebook with reality TV; and she talks of an interviewee of hers who implicitly related his not using Facebook to his not being 'a reality TV person' (p. 1050).

In addition to this, we might observe that both reality TV and social media focus on the ordinary, the everyday and the mundane (interestingly, Paddy Scannell (2009) has described television in the 1950s as the 'medium of everyday life itself' (p. 208)): online this is reflected in the quantity of phatic communication (Miller, 2008). More generally, they are both demotic and feature ordinary people (see Turner, 2010, whose book includes chapters on both reality TV and SNSs).

Conclusion

'Sharing' today is so powerful a concept because it encapsulates the therapeutic ethos *and* the centrality of social media in our everyday lives. But more than that, the word 'sharing' directs us to the role of social media in the presentations of authenticity that precisely help to define the self in the therapeutic era. The fact that social media firms make use of 'sharing' as a marketing tool does not mean that what we do online is not really sharing, but rather unsubtly shows us that our interpersonal relations – including our romantic ones (Illouz, 1997) – do not exist outside of a political economy. If ours is the age of sharing, it is because in 'sharing' converge, on the one hand, the authentic expression of self both as a means of intimate relationship construction and maintenance and as a means of self-understanding, and, on the other, the mode of our participation in social media.

This implies that while calling our participation in social media 'sharing' certainly serves companies such as Facebook and Instagram, their use of the term is not (only) a crass marketing ploy aimed at playing on our heartstrings. In this chapter I have drawn a line from the period in time when 'sharing' came to mean 'communicating' – almost a hundred years ago – through to our presentations of self on social media today. This is not with the intention of then saying that what we do online is 'really' sharing. It is, though, to point to the important cultural similarities between 'sharing' in the Oxford Group and updating a status, similarities that are rooted in therapy culture. Samuel Mateus's (2010) use of the Lacanian concept of 'extimacy' is useful here. The idea is that the subject 'only reaches his inner self by making it public and sharing it. It is extimacy's condition that enables the very possibility of individuality' (Mateus, 2010: 69). You become who you are by presenting – sharing – your self to (or with) others.

Alison Hearn's (2010) discussion of 'public intimacy' also reminds us that this sharing is highly commodified and shows how authentic expression is turned into money, unlike in the Oxford Group. But my point is that the 'original' version of sharing as a type of public intimacy – taking the Oxford Group as an exemplar of this – was also performed in a context that, if not itself commodified, was part of a society that was becoming ever more commodified. For scholars such as Andrejevic (2004) and Lears (1983), this is crucial for understanding the emergence of the practice of sharing-as-telling in the first place.

In addition to pointing out the structural affiliation between sharing in the Oxford Group – representing here the therapeutic ethos – and sharing on social media (authentic speech about the self with the aim of strengthening or healing the self and sustaining relationships with others, for instance), I shall return to the observation, mentioned at the chapter's start, that sharing in therapy culture, as well as in the sharing economy, is about making public what is private. We can see this quite neatly by looking at criticisms of practices of sharing-as-telling from different times. For instance, the guests on *Oprah* and heavy users of Twitter have all been accused of over-sharing and of polluting the public sphere with either inanities or downright inappropriate content (for a discussion of this criticism regarding the use of Twitter, see Arceneaux and Weiss, 2010; for a treatment of such criticism regarding Oprah Winfrey, see Illouz, 2003, especially Chapter 8). In both instances, the essence of the critique is that the boundary between the private and the public has not been maintained, and that information that should properly be kept private is actually being made public. We can even see a similar critique directed at the Oxford Group: recall its critic who panned the Group for its 'over-emphasis of the emotional side of religion, in the unrestrained public confession of sins' (Clark, 1951: 167). In therapy culture, sharing – whether online or off – expands the public at the expense of the private.

Moreover, this is a process that has come to be increasingly mediated by for-profit institutions – TV production companies, the social media industry – which is a recurrent theme throughout this book. Lears (1983) helps us understand the cultural and economic context of 'sharing' in the Oxford Group, but although the Oxford Group obviously had an interest in attracting members, this interest was not financial. Today, sharing on Airbnb, Facebook or *Oprah* contributes to the wealth of private corporations. Tracing the therapeutic sense of 'sharing' – from the Oxford Group, through AA, reality TV and up to social media – shows once more that the most pressing matter is not whether it is 'really' sharing, but rather to highlight the shifts in the socio-economic context of this sharing, and the increase in power of commercial organizations in mediating it.

6
Sharing Files

This chapter is about file sharing. There are a number of good reasons to discuss file sharing in a book about sharing in the digital age. One is that it presents us with an opportunity to see how the theoretical framework for understanding sharing, which has been developed in this book, can be implemented in relation to a specified sharing practice. Another is that despite the rise of streaming, file sharing is still one of the main uses of the internet.[1] A third reason is that it allows us to analyse the incentives for sharing and the sanctions for not sharing within a well-defined field, with the expectation that such an analysis will shed light on other sharing enterprises as well.

There are a number of ways to enforce norms regarding sharing (understood here primarily – but not only – in its distributive sense). The most common are social – think of kindergarten teachers' admonishments of children who are not 'sharing nicely'. When we get a bit older, norms of sharing are enforced by variations on tit-for-tat rules: if you don't share with me, I won't share with you (Kennedy, 2016 discusses the significance of reciprocity in sharing). Sometimes norms of sharing are deeply embedded in cultural practice. The anthropological literature on hunter-gatherer societies, for instance, strives to show this. But we can see norms of sharing at play in our society through a simple thought experiment: imagine replying to a neighbour who had just asked for a couple of eggs, or a cup of sugar, that although you have those ingredients in your home, you think people should be self-sufficient and that they should learn to ensure they have everything they need. That this is ridiculous says something (admittedly, something fairly obvious) about how norms of sharing are sometimes simply unquestioned.

Sometimes, though, the means by which sharing is enforced are technological. For example, Skype works by creating ad hoc networks of users, all of whom contribute bandwidth, mostly unwittingly. In other words, when using Skype, we are sharing our bandwidth with other users, and we have no alternative but to do so. Waze, the social GPS app, works in a similar fashion. Drivers can benefit from knowing in real time where there are traffic jams on their journey, but only by allowing Waze to know where they are and how fast they are driving as well. This chapter discusses various technological means for enforcing sharing in the context of private file-sharing communities. We will note, however, that even technological solutions to ensuring fair sharing raise normative discussions and new, unexpected problems. It should also be noted that while file sharing is a contentious practice, in the following I am agnostic as to its ethics; in other words, I am not interested in critiquing (or endorsing) file sharers from a position external to their practices. I am interested, however, in the ways in which the notion of 'sharing' frames those practices on the one hand, and the manner in which certain file sharers talk about the ethics of different models of file sharing on the other.

The chapter is composed of two parts. The first includes a discussion of the very term 'file sharing', arguing that it is an apt term, and certainly preferable to 'piracy', the metaphor that governments and the entertainment industry would prefer to see used. The second lifts back the curtain on a niche aspect of file sharing, namely, private BitTorrent file-sharing communities. These communities are of particular interest because of the discussions their members hold about how one should share, when sharing is too much sharing, incentives and sanctions aimed at producing the right amount of sharing, community and altruism, and the difference between sharing and capitalist economies.

In both parts of the chapter, I shift the focus of attention from the 'file' part of file sharing – which leads to discussions about ownership of the file, the ease of duplicability of the file and so on – to the 'sharing' part. As I hope to show, this opens up new ways of talking about file sharing.

Why File Sharing is Called File Sharing[2]

File sharing is contentious. One aspect of the debates over this contentious practice, or set of practices, concerns its terminology: on the one hand, we have 'piracy' and, on the other, 'file sharing', two metaphorical terms with quite different associations. Given that 'structures of metaphorical meaning shape our understanding' (Billig and MacMillan, 2005: 461), efforts to impose the metaphor of piracy on the digital distribution of copyrighted media are perhaps not surprising (for incisive critiques of these attempts, see Gillespie, 2009; Mirghani, 2011; Yar, 2008). Here, though, I am interested in the term 'file sharing', which I conceptualize as a linguistic and conceptual hinge that both locks together, and enables movement between, technical, discursive and institutional dimensions of contemporary digital culture.

Two questions require our attention here. First, why is file sharing called 'file sharing'? In answering this question I trace the use of the word 'sharing' in the context of computing back to the large time-sharing systems of the 1960s through to the prevalent meaning of file sharing today. In doing so, I show how the object of sharing has shifted from hardware to data in parallel with a shift from scarcity to abundance. The second question entails enquiring into the rhetorical force of the term 'file sharing'. Here, I argue that the notion of file sharing reverberates with positive appraisals of sharing as a valued social practice, as described earlier in the book.

In the background of this discussion, and very much like in the case of the sharing economy, we find people who take issue with whether file sharing is 'really' about sharing at all. 'It isn't sharing,' said Richard Parsons, the CEO of Time-Warner, to the US Congress, 'it's online shoplifting' (cited in

Litman, 2004: 23, n.94). Or, in the words of a British music manager: 'I don't know why it is called filesharing. It is stealing music' (Andrews, 2009). The idea that file sharing is not really sharing can also be found in academic writings, such as those of economist Stan Liebowitz, who makes the point that '[I]ndividuals do not "share" the files that are moved back and forth on the Internet' (Liebowitz, 2006: 4). In a similar vein, in his book on the rhetoric of peer-to-peer debates, rhetorician John Logie (2006) argues that '[T]he uncritical maintenance of "sharing" as an overarching metaphor for peer-to-peer file transfers distorts the debate', urging advocates of peer-to-peer technologies to 'acknowledge that peer-to-peer transfers are not a form of "sharing" as the term is conventionally understood' (p. 100). To this, one might reply that nor are peer-to-peer transfers a form of 'piracy' as the term is conventionally understood, except, of course, for the fact that the digital transfer of media files is by now precisely the conventional (or at the very least, *a* conventional) understanding of both 'sharing' and 'piracy'.

It is also pertinent to any discussion of the term 'file sharing' to note that the practices it describes are not new, and nor are the legal responses to them: indeed, the first copyright statute – the Statute of Anne – was introduced in England over 300 years ago, and even in the past few decades public, legal and academic debates have been ignited by the photocopier, the video-cassette recorder, and the dual-cassette deck, to name but three technologies for reproduction. However, the practices discussed in those contexts are *not* described as sharing. A full review of the literature covering those topics is beyond the scope of this book. However, if we take a smattering of the more important texts on copyright and the law, where the practices we today call 'sharing' are discussed in great detail, the term 'sharing' is at no point used in descriptions of what people are actually doing. (These texts include: J. Boyle, 1996; Breyer, 1970; Kaplan, 1967; Ladd, 1982; Laing, 1985; Nimmer, 1969; Patterson, 1968; Plumleigh, 1989. Nor does the famous case of *Sony Corp.* v. *Universal City Studios* contain any references to sharing.) This suggests that the origins of the term 'file sharing' lie in the *digital* nature of the practice, and that we should thus locate it within the history of computing.

An extremely brief history of computing

The first electronic computers were built during the Second World War by scientists in the US, England and Germany. These computers could only do one thing at a time, and users had to form an orderly queue in order to gain access to their computing power. However, towards the end of the 1950s, the idea of granting access to a computer to multiple users simultaneously began to be floated, a practice that came to be called *time sharing* (Hauben and Hauben, 1997: Chapter 6). The sense of 'sharing' here was quite literal: the mainframe computer's time was divided up between the users. Even then, however, the positive connotations of the term were not lost on computer scientists. For instance, in 1955, a group of IBM customers formed an association of computing professionals aimed at preventing duplicated programming efforts. The group called itself SHARE, a word that its first Secretary said 'describes very well the objectives of the group' (Mapstone and Bernstein, 1980: 363). The user group's name conveyed the values of 'cooperation and communication' (Armer, 1956: 2), as well as voluntarism and collaboration (Akera, 2001), and reflected an important aspect of the early computing industry (for two excellent histories of computing, see Campbell-Kelly and Aspray, 1996; Ceruzzi, 1998).

The development of the integrated circuit and the microprocessor began to make it feasible to move computing power from the mainframe to the user's terminal device, or alternatively to build smaller and cheaper but more powerful mini-computers, a trend that accelerated during the 1970s and that was expressed in the adoption of the microcomputer, or personal computer, as we know it today. The proliferation of computers created a new challenge, namely, how to access data on other computers. The solution was to provide remote access to another machine's disk drive. This was known as *disk sharing*, or *shared disk access*.

In terms of the logics of sharing, this represented a shift from sharing as dividing to a sense of sharing as 'having in common' or, in other words, a shift from 'sharing' to 'shared', which we can understand as a move away from the zero-sum game of sharing understood as divvying up a limited entity (such as a computer's time). Second, the move from time sharing to shared resources also reflects the increasing importance of computers as machines that deal with *information*, rather than as machines that merely compute. Users did not need access to remote disk drives because of their computing power, but rather because they contained useful data.

Once access to remote machines was possible, so was *file sharing*. However, we should distinguish between file sharing as providing access to shared files, in the sense of being held in common, and file sharing as the reproduction and distribution of files. When I set up printer sharing on my home network, this is the former sense. However, it is the latter sense that we now associate with the term 'file sharing' – downloading (copyrighted) films, music albums, books, etc. – a practice that attained mass popularity with Napster in 1999, which was shut down in 2001 following a law suit served by the Recording Industry Association of America. Indeed, some see Napster as directly responsible for bringing the term 'file sharing' into the public's consciousness (Liebowitz, 2006).

By briefly discussing the meanings of sharing in the context of computing over the last six decades, I hope to have offered something of an answer as to why file sharing is called 'file sharing', as well as indicating a shift in the usage of the notion of sharing in that context. Sharing has always been part of the terminology of computing. Specifically, the meaning of sharing in the context of file sharing would appear to be an instantiation of sharing as accessing the commons: this we saw in the contexts of time sharing, disk sharing, shared resources and file sharing when understood as making files accessible to remote users (though without making copies of them). File sharing is very similar to the model of the commons in that, at least with some forms of file sharing, such as that represented by Mega (the successor to Megaupload), users put files in a repository that others have access to. It is unlike the commons, though, in that, as mentioned above, downloaders take nothing away from the commons by downloading a file: in this commons, there can be no tragedy (Hardin, 1968).[3]

Given the long history of sharing – in various senses – in computing, calling file sharing 'file sharing' was a natural step. This differentiates file sharing from piracy in quite a significant manner: piracy is a metaphor selected by the powerful and imposed upon the weak (Van Dijk, 1993; Wodak and Meyer, 2001), as highlighted by Gillespie's (2009) and Yar's (2008) articles on anti-piracy campaigns targeted at schoolchildren; 'file sharing', on the other hand, is a term that has emerged bottom-up from the field. Moreover, it did so in the context of an industry in which, at least until the late 1960s, there was no software market as there is today: software was developed in-house by the user or came bundled with the hardware; the concept of purchasing software thus actually post-dates the practice of sharing it.

What we have learnt from this brief history is that: (1) the use of the word 'sharing' in this context has become more metaphorical and moved further away from its original sense of division into parts; (2) the context of sharing in computing has shifted dramatically from one of scarcity of computing resources to one of abundance;[4] and that (3) the object of sharing has shifted from hardware to information, corresponding with the evolution of computers from computing machines to devices that produce, analyse and store information. This is a shift that echoes the changes in the uses of 'sharing' in social media described in Chapter 3.

The (unintentional) rhetoric of 'file sharing'

We have already seen how sharing is a practice almost universally approved of, and how it bestows a positive veneer on that which it describes. This all-round sense of goodness is certainly relevant in the case of file sharing (as it is regarding the sharing economy too), and it is undoubtedly an aspect that frightens its opponents, hence the plethora of government and entertainment industry representatives insisting that 'it isn't really sharing'. As we saw in the previous chapter, sharing is a central component of our therapeutic culture, and the communicative aspects of sharing are also enacted by the term 'file sharing', along with its distributive aspects. This is suggestively captured in Liebowitz's (2006) critique of the activity and terminology of file sharing. 'File sharing' is a misnomer, he argues, because file sharers 'do not experience these files together nor are they likely to ever meet or even know one another' (p. 4). It is not sharing because it is impersonal, argues Liebowitz, suggesting that sharing entails familiarity between the sharers.

Those who prefer to talk in terms of piracy are critical of the term 'file sharing' because what it describes are not zero-sum acts of division, and because what we might call the rainbow effect of sharing does not appear to hold in this case.[5] This, however, is to ignore the communicative aspects of 'sharing', and in particular the idea that sharing our feelings and emotions is how we sustain social relationships. Of course, this is not to say that file sharers feel close or connected to one another (though they may: see Beekhuyzen and von Hellens, 2009; Cenite, Wang, Peiwen and Chan, 2009, on the sense of community among some file sharers). Rather, my point concerns the rhetoric of file sharing: as well as drawing on sharing tangible things as a prosocial type of behaviour, it also brings to mind a non-tangible type of sharing, namely that of emotions and feelings, which is central to contemporary interpersonal sociability. We shall now see how file sharers themselves draw on this sense of sharing in their debates over how much sharing is the right amount.

It Ain't What You Share (It's the Way That You Share It)

We now turn our attention to the practice of file sharing, rather than its nomenclature, if only for the purposes of reuniting them again later on. The remainder of this chapter is about file sharing that uses the BitTorrent protocol. Specifically, it is about private, members-only websites that provide links for content that is downloadable using the BitTorrent protocol, and that monitor whether their members are sharing with others as much as they are downloading for themselves. Drilling down even further, this section is about the ways that members of private BitTorrent sites talk about the mechanisms used to monitor the extent to which they are sharing files in a reciprocal fashion or not.

There are a number of good reasons for looking at these private file-sharing communities. First, they constitute a phenomenon that, in the context of research into file sharing, has been relatively under-studied. This is partly to do with the growing maturity of research into file sharing; it is arriving at a stage where we should be alert to distinctions that perhaps a few years ago were less important to demarcate. The growing body of studies into private BitTorrent sites, though, unequivocally shows that they have idiosyncratic features that should be teased out. For instance, we can ask whether the various mechanisms for requiring members to upload as much as they download have unintended consequences (spoiler alert: they do) and what implications this has for the widely held belief that, when sharing, one should not only take, but give as well. Relatedly, there is a twist, as it turns out that in closed file-sharing communities, the commodity being shared is not files at all, but rather a proxy for currency in those communities, namely *share ratio*.

Second, as suggested at the start of the chapter, private BitTorrent communities offer us a contemporary site for examining the sticks and carrots that are used to cajole and persuade people to share fairly. As a sociotechnical system, BitTorrenting is pretty new (the protocol was released in 2002), but the issues that arise in discussions of it draw on time-honoured topics of debate over any kind of commons. In this regard, by looking at file sharing we are able to engage with far more general questions about sharing. For instance, how are members of a sharing venture to be encouraged to share equitably? How is 'sharing equitably' defined? And, what sanctions might there be against non-sharing and free riding? These questions are akin to those that every society or group, large or small, has to find a way of answering.

Third, file sharing offers us another sphere of action in which we can explore the place of the word 'sharing' in framing and constructing the activities in it. In other words, it is a site in which we can see the metaphor of 'sharing' at work and enquire into its modus operandi there, both by looking at the ways in which the word itself is used, as well as the ways it is explicitly highlighted as a resource in attempts to shape the moral contours of the file-sharing endeavour.

Torrenting

There are currently two main forms of file sharing.[6] The first, which is of less interest to us from a sharing perspective, is represented by sites such as Kim Dotcom's Mega.com. For these, people upload files to a central repository from which other users can download. As far as downloaders are concerned, they simply click on a link and wait for the file to arrive on their computer. This form of file sharing clearly places very high demands on the central repository in terms of its use of bandwidth as everyone downloading the file is accessing the same server.[7] It also offers law-enforcement agencies a single target at which to aim.

The second main form of file sharing uses the BitTorrent technology. BitTorrent differs from the former model in three main and related ways. First, the file is divided into chunks that are downloaded separately and recompiled by the user's BitTorrent client software. Second, the user does not download the whole file from a single centralized source, but rather from other users with whom the downloader is connected in an ad hoc, peer-to-peer *swarm*. Third, and crucially, while downloading a file, the user is also uploading it to other people in the swarm. This is related to the first difference just mentioned: because the file is divided into chunks, which can be downloaded in any order, different users will have different chunks on their computers at any given time. So User A may have chunks 1, 3 and 5 on her computer, while User B has chunks 2 and 4. At this point – and this is clearly an oversimplification – User A will be downloading the chunks she is missing from User B, who in turn is downloading his missing chunks from User A. This is a peer-to-peer network where sharing is built in: you cannot download a BitTorrent without making the chunks you have already downloaded available for upload by other users.[8]

One of the outcomes of the BitTorrent technology is that it reduces the phenomenon of free riding, which is a problem faced by sharing ventures, be they contemporary or historical.[9] Indeed, the problem of free riding is what drove Garrett Hardin to pronounce that the end of the commons was tragedy (Hardin, 1968). As noted, the way this problem is reduced is by having downloaders upload to other users parts of the file that they have already downloaded. Users cannot simply download the entire file without making any contribution whatsoever to other people who are also trying to download it. If there is a politics to the artefact of BitTorrent (Winner, 1980), or if the technology implies any values, it is that one should not take without giving back; and, ideally, one should give back more or less the same as one has taken.

At this point, we must add the observation that the vast majority of us enjoy faster internet download speeds than upload speeds. If my download speed is five times faster than my upload speed, then in the time it takes me to download a torrent, I will only have succeeded in uploading 20 per cent of it back to other users. However, if I leave the swarm the moment I have finished downloading, I am reducing the number of users from whom other downloaders can access the chunks they still need. If everyone did this, then there would be users unable to finish downloading the torrent at all. Given the logic of BitTorrents, then, and the nature of our internet connections (faster download than upload), the right thing to do, from the perspective of sharing fairly that is implied by the BitTorrent technology, is to continue to make all the torrents one has finished downloading available for other users – this is known as *seeding* them; downloading is called *leeching*.[10] On the face of it, the good file sharer will ensure that she uploads at least as many bytes as she downloads, information that is readily supplied by BitTorrent downloading software. Put differently, what the BitTorrent protocol implies, normatively speaking, is that users should aim to maintain a *share ratio* of at least 1, where share ratio is the quantity of data one has uploaded divided by the quantity of data that one has downloaded: thus, if I download a 100 GB file and upload 50 GB of it to other users, my share ratio will be 0.5; if I seed for long enough, and upload 200 GB, my share ratio will be 2. Clearly, users who arrive at the swarm after I have finished downloading the file are counting on me (and others) leaving my BitTorrent software open so they can download it. In fact, it would seem quite sensible to say that the right thing to do, when sharing torrents, is to seed as much content for as long as possible, thereby making more content available to more users for a longer period of time.

As mentioned, if everyone simply downloaded the file and closed their BitTorrent clients (a practice known as *hit-and-running*, often abbreviated as HnR), then the torrent would die and would not be available for downloading by others. This kind of behaviour is impossible to prevent in what are known as public or open torrent trackers: peers are anonymous to one another; one's history is unavailable to other users (meaning that you cannot know if I am a serial hit-and-runner or, in other words, one has no reputation to uphold); and, in any case, even if you find my behaviour repugnant, you cannot direct your BitTorrent software not to connect to me. Because keeping a torrent alive requires that individual users allocate disk space for the file and bandwidth for uploading it to other users, and because there are no consequences to the hit-and-runner for hit-and-running (insofar as no one knows you did it), torrents tend to have a short life: 40 per cent of them are dead within a month (Kaune et al., 2010). It was partly in order to overcome this problem that private trackers emerged.

Private file-sharing communities, or *private trackers*, are closed online groups. They are often focused on a particular kind of content, such as music, or a genre of music, TV shows or cultural content from a particular country. What differentiates private trackers from public BitTorrent sites is that they keep a record of the quantity of data uploaded and downloaded by each member; that is, they know each member's share ratio, whether they have brought new content into the community by uploading fresh torrents, whether they are seeding any of the torrents that they have downloaded or whether they are hit-and-runners, and how fast they upload and download data. The main parameters on which status in private trackers is based are the share ratio, and the amount of data uploaded to other users. Different sites have different policies concerning their members' share ratio, but they (nearly) all insist that members keep their share ratio above a certain threshold.[11] Systems for regulating users' share ratio are known as *share ratio enforcement* (SRE), defined by Paul Aitken in his ethnography of BitTorrent sites as 'the single greatest structural and organisational factor that differentiates private sites from the much more egalitarian logic of public BitTorrent filesharing' (Aitken, 2012: 96).[12]

In the remainder of this chapter we shall learn what members of private tracker sites think about SRE and some of the unintended consequences that different mechanisms entail, and we shall see how associations with 'sharing' are expressed through users' comments on different sites' share ratio enforcement systems. Moreover, given that these communities are dedicated to file sharing, we shall ask how concepts of what constitutes sharing fairly are debated. Based on an analysis of forum discussions about the best (most moral, most efficient, most just) way to share files in private filesharing communities, it would seem that file sharers face up to a range of complex ethical issues that involve distributive justice, game theory, capitalist vs socialist models of the economy and more.

It should be noted that these are more than just casual downloaders. Most private trackers are not easy to gain access to – Aitken writes in some detail about the labour one must invest before attaining membership to one, describing it as 'a complex and involved process that requires a significant investment of time, a considerable level of technical competency, and enough "web savvy" to know where and when to look for information regarding private sites' (Aitken, 2012: 93). Also, these users feel strongly enough about file sharing to comment in a forum thread; that is, they are interested enough in file sharing to be spending time not only reading forum posts but also writing their own contributions.

Before delving into the forum posts themselves, though, let us get a better understanding of how private trackers and their share ratio enforcement mechanisms operate.

Private BitTorrent trackers

Computer scientists have shown much interest in the share ratio enforcement systems adopted by various private trackers, studying them mostly from the perspective of network efficiency. These studies are able to provide us with some important background information. As we shall see, though, they can only take us so far.

The introduction of mechanisms to incentivize/force users to seed appears to have worked, at least according to computer scientists: private trackers do offer better download speeds and their torrents do live for longer (Chen, Chu and Li, 2012). However, some fascinating unintended consequences and peculiar paradoxes have also emerged. The first unintended consequence is that SRE creates a problem that is the mirror image of that which it was intended to solve. If insisting that users maintain a certain share ratio overcomes the problem of users' being minimally motivated to upload, when share ratio becomes a kind of currency it creates the problem of 'poor downloading motivation' (Chen, Chu and Li, 2012). Users become cautious about downloading because they need to protect their share ratio; however, because every bit downloaded by one user is a bit uploaded by another, users wishing to upload (and improve their share ratio) are faced with other users who are reluctant to download (Jia, Rahman, Vinkó, Pouwelse and Epema, 2011).

This problem is particularly acute for users with a poor share ratio, as they cannot download files. They can of course upload new torrents, but even if they are leeched (downloaded), there is no guarantee that the uploader will be the primary beneficiary, even if the torrent is downloaded multiple times: once the torrent has been leeched in its entirety by another user, another key feature of the BitTorrent protocol kicks in, namely, its preference to connect leechers (downloaders) with the fastest seeders (uploaders). At this point, differences in the speed of users' internet connections come strongly into play: if my internet connection enables me to upload at a speed of 5 Mbps, but yours enables you to upload at a speed of 100 Mbps, and we are both in the same torrent swarm, the BitTorrent protocol determines that users will be directed to you for their download needs. Thus, your share ratio will rise while mine remains stagnant. As we shall see, this is an issue that taxes members of private trackers who ask themselves whether this state of affairs is fair, but before addressing that, we need to add another term to the mix: *seedboxes*.

A seedbox is space on a remote server that is set up to leech and seed torrents. When a user rents a seedbox,[13] he is renting server space on which to save his torrents and is getting very fast up- and download speeds, typically of 100 Mbps. (At this speed, you can download the contents of a DVD in about five minutes.) The seedbox user now has a huge advantage over regular home broadband users when it comes to accumulating share ratio. Some users consider this an unfair advantage, as it makes it harder for them to seed and thus improve their share ratio, which in turn would enable them to download other torrents, which, after all, is the purpose of their being there in the first place.

As noted, the computer science literature helps us understand some important features of private file-sharing sites that operate share ratio enforcement systems: they are better than public trackers in that they offer faster download speeds and tend to keep torrents alive for longer; they create an economy wherein share ratio is the currency; and they create a situation in which supply exceeds demand, which for certain users – particularly those with slower internet connections – can prove an obstacle in building up a big enough share ratio that will enable them to continue to download torrents (Chen, Chu and Li, 2012; Jia, Rahman, Vinkó Pouwelse and Epema, 2013; Kash, Lai, Zhang and Zohar, 2012; Liu, Dhungel, Wu, Zhang and Ross, 2010).

However, while this background knowledge is essential, it does not contribute to our understanding of what the actors themselves think about these systems and how they conceptualize the fair and just way for things to be regulated. Perhaps this criticism can be brushed off by asserting that it is irrelevant: articles about the efficacy of this or that share ratio enforcement scheme do not claim to contribute to our understanding of how users understand the system. However, they do make assumptions about the members of private trackers that do not appear to hold water, in particular that we are dealing with rational actors in game theoretical situations. For example, an article about incentives for sharing in Napster states its assumptions 'that agents are economically rational, and that they act to maximize their expected utility' (Golle, Leyton-Brown and Mironov, 2001: 265). While there clearly are file sharers who take a utility-maximizing view of file sharing – and these are criticized by certain other file sharers – there are others who, according to researchers, are operating within the sphere of 'commoning' (Caraway,

2012) and gifting (Giesler, 2006; Giesler and Pohlmann, 2003; Ripeanu, Mowbray, Andrade and Lima, 2006), or at least of reciprocity (Cenite et al., 2009). Similarly, findings from Beekhuyzen's study of private file-sharing communities (Beekhuyzen and von Hellens, 2009; Beekhuyzen, von Hellens and Nielsen, 2011) suggest that a contrarian standpoint vis-à-vis the music industry is at least part of what motivates file sharers. Similarly, while Andersson Schwarz urges us to beware of attributing anti-capitalist motivations to all file sharers, arguing that file sharing 'thrives on the same capitalist system of cultural exchange that it forms a part of' (Schwarz, 2014: 163; see also Tetzlaff, 2000), there are certainly those for whom sharing is an anti-establishment practice (see Guadamuz, 2002, on what he calls the new ethic of sharing in cyberspace).

So, rather than assume that the actors are rational and seeking their highest individual utility, let us see what they think about what they are doing, and let us try to understand why it ain't what you share, it's the way that you share it.

File sharers writing about file sharing

I now turn to a presentation of debates about SRE and the proper way to share files as carried out in three online forums: one from a well-known private music file-sharing website (Forum 1), another from an open forum dedicated to discussion of torrents and private trackers in general (Forum 2),[14] and the third from a private site also dedicated to talking about the torrent scene (Forum 3).[15] From Forums 1 and 2, I analyse two lengthy threads: one has 114 posts contributed by thirty-nine users, and the other has 146 posts contributed by fifty-five users, giving a total of 260 posts by ninety-four users. From Forum 3, I analyse six different threads in which seedboxes and share ratios are discussed. These threads comprise a total of 194 posts written by around 120 different users. In addition, I also draw on the Research Bay project, a survey of 70,000 Pirate Bay[16] users (Larsson et al., 2012). The survey included an open question in which respondents could write freely about file sharing.[17] These answers are searchable, and I here make use of the thirty-four answers in which participants used the word 'ratio'.[18] All texts were imported into MAXQDA, software for qualitative analysis, where they were coded.

The following analysis does not make claims about the relative prevalence of certain points of view; rather, it treats the very expression of a point of view – even if only once by one user – as indicative of a set of cultural beliefs. Having said that, it is notable that the kinds of arguments made are quite similar to those presented in earlier studies of file sharing (see Aitken, 2012; Giesler and Pohlmann, 2003; Tetzlaff, 2000). The fresh angle that I advance in my reading of these texts is that the notion of 'sharing' as implying an egalitarian and reciprocal mode of economic relations reverberates throughout the discussions.

File sharing, sharing and community

While the forum posts I read cover a range of debates about SRE and private torrent sites, for the purposes of this chapter I focus on those posts that draw explicitly on the notion of sharing. For many of the forum posters, 'sharing' is an important resource. Indeed, certain users explicitly frame the debates in the forums in terms of sharing and the associated values of fairness and caring. As we shall see, an interesting consequence of discussing share ratio enforcement mechanisms and overly inflated share ratios in terms of sharing is that the focus shifts from the individual to the community. The argument I am making is not a straightforward causal one; it is not simply that because we are dealing with file sharing and share ratios people have notions of fairness and community in their heads. Rather, for file sharers who believe in the values of 'sharing' (whatever they are), file sharing is a community endeavour, and the welfare of the community plays a large part in how they conceive of the fairest division of labour. As we shall also see, a commitment to the community does not necessarily translate into an entirely lax approach to share ratios, and later in the following discussion we shall encounter users who suggest – quite politely, for the most part – that if there are users who cannot meet the demands required by the community, then perhaps their file-sharing needs might be better met elsewhere.

References to sharing as a positive activity are plentiful in the forums. A contributor to Forum 1 says that 'sharing is what it's all about'. Another writes, 'I like to share my stuff', and yet another, 'I'm here to share'. While we shall come on to the creative solutions to the problems created by SRE very shortly, let us note now that a poster to Forum 2 promotes their solution by saying: 'This would allow everyone to do their fair share, while still allowing the high speed seeders [to] do their thing! In the end, everyone who tries would have a good ratio and can share.' Here, sharing is related to fairness, through the idiom of a 'fair share', while defining the constitutive activity of file-sharing communities as sharing ('everyone [...] can share'). That same user also defines being part of a private community as 'based on sharing and doing your part'. Posters to Forum 3 also talk about sharing. As part of one of the debates that we shall delve into shortly, a contributor describes what s/he thinks that 'Any tracker that cares about its users' should do; otherwise, s/he says, 'switch to another protocol if sharing and caring isn't the goal', implying that 'sharing and caring' *is* the goal. Another poster responded excitedly to a how-to guide for beginners with the comment that 'if our filesharing community is still about sharing and caring' there is no better way to show it than by helping out newcomers. In yet another post that is also part of a debate over how best to organize private trackers, the poster writes that s/he 'Didn't realize seeding as much as possible would be perceived as a bad thing', before adding, 'Sharing is caring'. Once more we are shown how culturally accessible this rhyming aphorism is, and how it helps to frame people's moral attitudes.

These are not the only instances of posts whose authors are invested in the notion of sharing. One quite angry poster started a thread about overseeding. Overseeding is when a user continues to seed a torrent even though they have reached a share ratio of 1 for that torrent, and when by seeding it they are preventing other users from seeding it who perhaps have a greater need to upload and improve their share ratio – this would particularly apply to seedbox users, whose upload speeds are very high. The user wrote: 'I thought filesharing and being called a pirate meant something.... It was about sharing, not who has the best stats or forcing members into spending even more money to rather buy upload credit or pay for other means to download.' He later adds: 'Filesharing is about sharing, BT [BitTorrent] is about giving back what you received, lets [*sic*] work together to keep it that way.' Another user, who declares that s/he is 'poor as fuck', comments that they were attracted to pirating because of the Marxist idea that 'people share what they can afford and through that everyone wins'. Another defines 'what the torrent system is all about' as 'sharing with others like you'.

Sharing is contrasted with buying and selling (Belk, 2010, 2014), and file sharing, some believe, should not be an endeavour that costs money. However, because of the presence of seedboxes, states one poster to Forum 3, 'pirating is becoming a "pay" service in one way or another, as you are forced to spend money in some way to compensate for not being able to compete with seedboxes and super-fast connections for valuable upload'. Relatedly, the previous commenter unambiguously asserts: 'I refuse to pay for a service to better my chances on what was and always should have been free as filesharing was intented [*sic*].' Similarly, another contributor to the same discussion in Forum 3 says that 'The idea of having to get a seedbox in order to stay on a tracker is a bit off imo [in my opinion], it goes against pretty much everything I stand for', adding that 'it's not right to have to force members to spend their money on seedboxes in order to keep their accounts', while another contributor thinks that 'just because you can't afford a seedbox or good connection doesn't mean your ratio should hurt'. In response to a few comments that convey the message that everyone should just look out for themselves, a poster sarcastically writes, 'Nice to know that everyone is so caring'.

An even more explicit critique of the entrance of money and financial concerns into file sharing is found in Forum 3, where a user (the one who described himself as 'poor as fuck') asks, 'Didn't we all start sailing the high seas in order to avoid the trappings of capitalism to some degree?' In a similar vein, another poster to Forum 3 wishes more people would put the community ahead of the individual: 'People only care how fast they can get stuff for THEMSELVES, the other members are an afterthought. [...] The spirit of filesharing destroyed right there.' Also pushing a more collectivist approach, another poster states that 'it is your responsibility *as a pirate* to make room for others, if the Seeder-Leecher ratio hits 3:1 and you have a decent connection, just drop off the swarm and let others have a go at it' [emphasis added]. The phrase 'as a pirate' is revealing, as it suggests that pirates behave differently from everyone else, or that there are ways of doing things that make pirates stand out from the rest, and that one of the things that pirates do is unselfishly 'make room for others' who are also trying to better their share ratio.

Interestingly, the notion of 'sharing' is also used by some of those who have little time for users who complain about having a low share ratio: for instance, one poster asserts that they 'don't want to share, they just want to take'. Regardless of the fact that they *do* want to share, but find themselves unable to because of the predominance of seeders over leechers in private trackers, this poster nonetheless conceives of sharing as reciprocal; for him/her, just taking contradicts the values and purpose of file sharing.

Solutions to too much sharing

The forum posts analysed here include a plethora of solutions to the problem of overseeding, or of sharing too much. I shall not list them all (I counted over thirty-five distinct solutions and suggestions), but instead focus on those that have counter-intuitive implications for the sharing of files in that they urge file sharers to share less and slower.

One suggestion is for seedbox users to throttle their upload speeds so that they are more akin to home users. This would have the effect of removing the preference afforded by the BitTorrent protocol to very fast uploaders. Of course, it would also have the effect of making downloading slower for everyone else – a point frequently made in many of the responses to the use of seedboxes in the forums.

The second solution is for the ratio- and buffer-rich to download files that they do not really want, just so that they put their buffer back into circulation. A poster to Forum 1 says that s/he does this all the time: 'I download 50 random torrents a day, until I run out of hdd [hard disk drive] space.' Another says that in the past s/he 'randomly downloaded shit just for the sake of downloading it'.[19]

The third thing that posters to the forums suggest that users can do to level out share ratios among private-tracker site members is to stop seeding when there is no need to. This, it must be acknowledged, is quite a labour-intensive approach and requires that users analyse the state of the various torrents they are seeding in order to decide what to carry on seeding and which swarms to leave. There is something counter-intuitive about this; after all, it calls for file sharers *not* to share files – or at least, certain files and under specific conditions. Knowing how to share, it turns out, entails knowing when not to share. Interestingly, this is not just true of file sharing: in *Sharing the Journey*, Robert Wuthnow (1994) describes a woman who was ejected from a support group for talking too much. In that instance, her ratio of talking to listening was unbalanced, and the community rejected her.

Conclusion

In this chapter, we have seen why file sharing is called 'file sharing' as I located the emergence of the term in the short history of computing. Part of the resonance of the term today is clearly to do with the positive associations ignited by the word 'sharing'. However, inquiring into the origins of the term also sheds light on the shifts in computing from conditions of scarcity to conditions of abundance, and from a focus on hardware to a focus on information.

I next suggested three reasons for looking closely at file sharing on private trackers. The first was that we do not know enough about them. What the analysis presented here shows is that by enforcing a share ratio on their users, private trackers have brought about a number of unintended consequences into the world of file sharing. One is that, in private trackers, it does not matter *what* you share, but rather *how much* you share. In their efforts to increase their share ratio, some users download content they have no interest whatsoever in consuming just so they can upload it back to other users. Another is that some file sharers feel that other file sharers are sharing too much, and that sometimes the act of sharing – uploading content to other file sharers – is actually selfish and harmful to others. This leads us to conclude that in private BitTorrent communities, it is not content that is being shared, but rather share ratio, which has become a sort of currency. Some users clearly believe that this is a resource that should be distributed fairly. All agree that if there is a resource that needs to be managed in private BitTorrent sites, it is not content, but rather buffer;[20] this is a scarce commodity that some users hoard, while other users freely give it away to those they see as in need. Regardless of content, it is the methods for attaining and redistributing share ratio that are the focus of the discussions about how these sites should best be run.

The second reason for looking closely at private BitTorrent sites was that the sticks and carrots used in them might be of more general application to other sharing ventures. These sites have clear sanctions against breaking their rules: expulsion from the community. This is a pretty powerful stick to wave at file sharers. Incentives are also offered in the form of status within the community. Most of this status is symbolic (being called a Superuser, for instance), though sometimes concrete privileges are available too (special forums, early access to new torrents and more). In a way, the attribution of status in private file-sharing communities is similar to the reputation systems that operate on a great many enterprises within the sharing economy at large.

The third reason for studying these sites is to make explicit the place of the word 'sharing' in the arguments made for and against share ratio enforcement. From the excerpts presented above we can see that some users identify very strongly with the word 'sharing', arguing that 'that is what it is all about'. For some, it seems that they take their lead from the term, 'file sharing'. If it is called 'file sharing', they appear to think, then we should make sure that we are sharing. The easy associations between sharing and caring, and the notion of a fair share, also emerge. Others deploy the concept of sharing as they try to argue for a particular mechanism, usually one that involves the redistribution of share ratio in accordance to need, and not in proportion to how much has been paid for this or that seedbox. For these people, 'sharing' also expresses a form of ownership that counters that which is represented by the entertainment industry. Either way, 'sharing' is used as a concept that brings to mind 'caring' and fairness, very much as we saw earlier in previous chapters.

If file sharing was not called 'file sharing', these file sharers may well still have been doing it. However, the examples cited in this chapter show quite clearly how the word itself is influencing forms of thought and self-expression and setting expectations for the kind of interpersonal relationships that file sharers in private BitTorrent communities imagine they will find there. As Billig and MacMillan argue, 'structures of metaphorical meaning shape our understanding' (2005: 461).

7
Conclusion

'Sharing' has come to represent a range of spheres of our lives, and those spheres are converging under the term 'sharing'. By this I mean that a common set of values and associations with sharing informs our practices, and our reflexive interpretations of those practices, across the three social fields that I have analysed. The purpose of this book and the analyses it offers is not to have the final word in scholarship about social media, forms of intimate communication or the sharing economy. Rather, its purpose has been to explore a single word, 'sharing', and through that to say something about how we live, and hope to live, today. As I shall develop in these final pages, the notion of 'sharing' today involves the expansion of the public at the expense of the private in a manner that is increasingly mediated by digital, for-profit enterprises. I say 'involves' and not 'is', and I add the modulator 'increasingly' because 'sharing' is so rich in both semantic and practical terms that a single statement cannot capture its entirety.

'Sharing' is not just a homonym, but a polysemic homonym. The difference between the two is a function of whether the different meanings of the homonym are related. In the case of the word 'spring', for instance, they are not ('spring', the season, has nothing to do with 'spring', the tightly coiled piece of metal). In the case of 'sharing', though, they most certainly are. What links the different instantiations of sharing that have been discussed throughout this book is a set of values that we associate with 'sharing'. As stated in the Introduction and referenced repeatedly throughout, these values include openness, honesty, mutuality, equality, trust and more. This does not mean, though, that everywhere we find talk of sharing, we find these values. One of the things I have tried to show throughout is *when* 'sharing' historically came to be associated with the practices the word is used to describe: as we saw in Chapters 2 and 5, certain positive values came to be associated with sharing at specific social and cultural junctures (in particular the 1930s and the 1970s). This diachronic approach serves the function of reminding us not to essentialize sharing (or 'sharing'), or to ossify any particular one of its meanings. It also enables us to see that the current constellation of meanings around sharing is novel and far-reaching.

Meanings change over time. In some instances, the word 'sharing' was already in use and certain values became newly attached to it. This is the case with digital sharing: 'time sharing' and 'disk sharing' as used in the 1950s and 1960s were normatively neutral terms, and were close to the non-metaphorical and non-evaluative sense of sharing as dividing: when the plough shared the field, it was neither good nor bad. But as 'sharing' migrated to file sharing and particularly to social media sharing, it became infused with the values we now so readily associate with it. In Chapter 3, I suggested that this itself was a function of the construction of computers and computer networks in utopian terms. At other times, though, the values were already there (or were described as being there), which makes the application of 'sharing' seem natural. Examples of this are people saying that the internet has always been about sharing, or that ancient tribes were organized around sharing. In these different ways the metaphor of sharing shunts meanings around the different social fields that I have called spheres of sharing. Of course, these processes of meaning shift are messy, and different senses of the word – of any word – sometimes overlap. As linguist Eve Sweetser reminds us, 'if a word once meant A and now means B, we can be fairly certain that speakers did not just wake up and switch meanings on June 14, 1066. Rather, there was a stage when the word meant both A and B' (Sweetser, 1990: 9; cited in Larsson, 2013: 617).

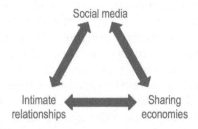

Figure 7.1   The spheres of sharing

Returning to the scheme presented in the Introduction, we can now say something about the arrows linking the semantic fields of sharing that have featured in this book (Figure 7.1). Social media and sharing economies are densely interrelated. Today's sharing economy is a high-tech digital economy – not least because of the digital connotations of the word 'sharing' – and we saw in Chapter 4 how online sharing is seen as at least partially responsible for a willingness to share stuff offline among Millennials in a striking example of how a metaphor is used to explain actual behaviour. Sharing economies and intimate relations are related in that the kind of interpersonal relations championed by sharing economies (both the for-profit and not-for-profit kind) is akin to that promoted by therapy culture: these are relationships based on honesty, trust, respect for the other and so on (recall the hugging encouraged at a sharing economy event held by OuiShare). As well as being digital, sharing economies – at least at the declarative level – strive to replace the functional, cold and impersonal relations engendered by capitalism with ties of trust and a sense of community and authenticity. Likewise, intimate ties are also seen as operating according to a different logic from that of the work- or marketplace. Finally, the connections between sharing on social media and sharing as a type of talk primarily subsist in the way that they both see the self and its relationship with others as based on authentic communication. The linkage here has also been made by John Durham Peters, who discusses 'the technical and therapeutic visions', both of which 'claim that the obstacles and troubles in human contact can be solved, whether by better technologies or better techniques of relating' (Peters,

1999: 29). Again, this is not to say that all social media communication is necessarily authentic and honest, nor that all communication with our significant others is of that nature either. It is to say, though, that in both cases the notion of sharing implies self-expression and the construction of the self through text (written and/or spoken).

As we have seen throughout this book, practices of sharing always straddle, and sometimes challenge, the constantly fluctuating and pervious divide between the public and the private. More accurately, but also more wordily, we should actually talk about that which is *perceived* as public and private. Even more wordily, but more accurately still, we should say that practices of sharing straddle the nebulous division between that which is perceived as properly belonging to the public sphere, and that which is perceived as properly belonging to the private sphere. I put it like this so as on the one hand not to essentialize the socially constructed character of the public and the private and, on the other hand, but relatedly, so as to emphasize their fluidity. In this regard sharing has directionality – from the private to the public – no matter whether we are talking about sharing as the constitutive activity of social media, sharing as a model for production and consumption or sharing as a type of talk.

It feels almost trite to say that online sharing expands the public at the expense of the private (on changes to the public/private divide in relation to the internet, see, for instance, Ford, 2011; Lasén and Gómez-Cruz, 2009). Public, after all, is the default.[1] On social media, and in other digital contexts, sharing means distributing, telling others (or everyone, as Hermida (2014) would have it), and certainly not keeping it to yourself. Extending the logic of reality TV, sharing on social media brings the mundane into public view (Marwick, 2012). In this regard, it really does not matter whether a Facebook profile is viewable by the entire world, or just by a user's Friends. In Chapter 5, we encountered the concept of 'over-sharing' – examples on oversharers.com include a photo posted by a mom of her kid's first poo in a toilet, and a description of sex in a stairwell; in short, over-sharing is when people publish 'embarrassingly intimate – or gross and disgusting – details' of their lives.[2] The implication of 'over-sharing' is that a normal amount of 'sharing' is fine.

Moreover, the very fact that the term 'over-sharing' was coined and widely adopted (it was Webster dictionary's 2008 Word of the Year) reflects how hard it is to talk about sharing in negative terms. In this regard, 'sharing' today is what 'community' was to Williams when writing *Keywords*: it is 'a word that is never used unfavorably' (Williams, 1976: 5).

As a model for production and consumption, the idea of sharing is closely linked to that of the commons. When someone shares their bread with a companion, they are transferring ownership of the bread from themselves to a kind of joint ownership. Indeed, in the field of collaborative consumption we can clearly see a reconfiguration of the public and the private: take Couchsurfing, for instance, which involves people inviting strangers into their home and letting them sleep on their sofa (for free); or the plethora of online lift-sharing schemes, whereby people allow strangers into the sanctum of their car. Without reigniting the debate discussed in Chapter 5, I acknowledge that there is a difference between letting someone make use of your stuff while you are not using it, and charging them to make use of your stuff while you are not using it. With the former, you are seeing a way of letting your stuff (or your time, or your whatever) contribute to the public sphere; with the latter, you are seeing a way of letting your stuff earn you money. The point I wish to make, though, is that in both instances you are letting other people, in many cases strangers, make use of your stuff. Relatedly, when a company goes public, it starts to sell shares. Sharing and publicness would seem to go some way back together.

Sharing as a category of speech also pertains to a shift in the boundary between the public and the private. In support groups, for example, where sharing is the constitutive activity, one speaks about one's private problems and places them in the public sphere (or at least that of the group). More generally, sharing as a form of communication between intimates is central to today's therapeutic culture, where the culturally accepted way of resolving conflict and solving problems is through talk. The point to be made here is that the contemporary intimate relationship – what Giddens calls the 'pure relationship' (Giddens, 1992) – is based on knowledge of the self, knowledge of the significant other's self and the imparting of that knowledge between the partners through talk. In this regard the self becomes less private. This is not to say that we make our selves entirely public, but through the type of speech we call 'sharing' we make more of our selves known to others, and we do so to a greater extent than would have been the case as little as fifty years ago.

A parody of the injunction to share – and an illustration of how the spheres of sharing are related – is found in Dave Eggers' novel, *The Circle* (2013). The book describes a Silicon Valley tech company (think Facebook plus Google on steroids) that we come to know through the experiences of a new employee, Mae, who rises quickly through the ranks. The company's Orwellian slogans are: 'Secrets are lies. Sharing is caring. Privacy is theft', and it is striving for a world where everything everyone does and, ultimately, thinks, is made public for everyone else to benefit from. Thus, when Mae, looking for a bit of solitude, takes a beautiful midnight canoe trip and does not post photos of it online, she is summoned by the boss, Eamon, to account for her actions. Later, in an interview-cum-chat on stage in front of the company's employees, Mae retells what she had told her boss:

'It was just selfish, Eamon. It was selfish and nothing more. The same way a child doesn't want to share her favorite toy. I understand that secrecy is part of, well, an aberrant behavior system. It comes from a bad place, not a place of light and generosity. And when you deprive your friends [. . .] of experiences like I had, you're basically stealing from them. You're depriving them of something they have a right to. Knowledge is a basic human right. Equal access to all possible human experiences is a basic human right.'

[. . .]

'You had a way of putting it that I'd like you to repeat.'

'Well, it's embarrassing, but I said that sharing is caring. [. . .] I think it's simple. If you care about your fellow human beings, you share what you know with them. You share what you see. You give them anything you can. If you care about their plight, their suffering, their curiosity, their right to learn and know anything the world contains, you share with them. You share what you have and what you see and what you know. To me, the logic there is undeniable.' (Eggers, 2013: 301–2)

In this extract, Mae deploys a range of meanings of sharing. To start, she compares not telling about her canoe trip to a child not sharing their favourite toy, eliding the communicative and distributive meanings of sharing. She does this again when she says that if you care about people, you share 'what you have and what you see' with them. Sharing instantiates your commitment to other people; failing to share, or worse, intentionally refraining from sharing, is a sign of selfishness and deviance.

In *The Circle*, sharing means technologically assisted radical publicness[3] within the context of a strictly capitalist system. The information Mae shares is harvested for its commercial worth, and Mae remains a lonely individual, alienated from her labour and her family. Despite sharing being equated with caring throughout Eggers' book, there is no suggestion that it heralds a return to communitas. Ultimately, the promise of sharing is violated. I think that what Eggers is describing is the realization of the 'Californian Ideology', described by Richard Barbrook and Andy Cameron as the outcome of 'a bizarre fusion of the cultural bohemianism of San Francisco with the hi-tech industries of Silicon Valley', and as a combination of 'the freewheeling spirit of the hippies and the entrepreneurial zeal of the yuppies' (Barbrook and Cameron, 1996: 44–5). This describes *The Circle* quite well, suggesting that we could posit Eggers' version of sharing as the slogan of the Californian Ideology: 'optimistic and emancipatory', yet blind to the 'polarisation of the society from which it was born' (Barbrook and Cameron, 1996: 62).

The annual festival of the Californian Ideology is Burning Man. The association between Burning Man and Silicon Valley, and Google in particular, has already been made plenty of times (two notable examples are Kozinets, 2002; Turner, 2009), and it is also a feature of the festival that attracts a great deal of press attention. Turner views Burning Man as an allegory of Silicon Valley culture, comparing the entire festival to the temporary nature of a high-tech product-development team. However, the association is not only allegorical, as business connections are made and work-related networks formed and strengthened; also, Google itself has made much of its mapping (satellite, 3D) of the festival, such that the company's association with Burning Man is part of its brand.[4] Turner's use of the concept of 'commons-based peer production' (Benkler, 2006) paints a picture of Burning Man as a sharing economy, albeit a temporary one. Indeed, argues Turner, its temporariness is exactly the point: it is by 'disowning the marketplace'[5] that Burning Man participants actually 'redeem its failures' (Turner, 2009: 91).

Robert Kozinets also grapples with 'inescapably porous boundaries between Burning Man and the market system' (Kozinets, 2002: 31). To start, he says that Burning Man 'provides people with the experience of living in a sharing, caring community, exemplifying the communal ethos said to be undermined by dominant market logics' (p. 33). By way of example, Kozinets tells of an informant who believed that the 'injunctions against commerce brought people closer together and thereby helped construct a caring, sharing sense of community' (p. 28). A central aspect of this is communicative: Kozinets talks about 'personalized interactions' (p. 29), 'self-expression' (p. 26) and the same 'therapeutic language' described by Lears (1995) in his work on the early American advertising industry. Indeed, sharing and caring feature strongly in Kozinets' ethnography of the festival. However, like Turner, Kozinets also develops a complex take on the festival's subversive potential. In fact, one might say that Kozinets' focus on consumption supplements Turner's attention to production. The catharsis of Burning Man 'ultimately props up the market system', says Kozinets (2002: 36), but at the same time it invites us ('some people, and sometimes') to think about 'emancipation and the role of markets and communities in our contemporary consumer culture' (p. 37).

Sharing at Burning Man captures much of what I want to demonstrate about the word today. It is countercultural and subversive; its associations with the high-tech scene are clear to see; and it promotes a mode of interpersonal relations that is based on authentic communication, equality, trust, acceptance of the other and mutuality. And yet. In 2014 the festival gained a degree of infamy as the presence of luxury camps with paid staff became impossible to ignore. Contravening the festival's principles of radical inclusion, radical self-reliance and decommodification, private luxury-camp organizers were selling tickets for $20,000 and promising a fully catered, air-conditioned experience for paying, wristband-wearing customers.[6] Following a backlash, the organizers changed the rules about 'turnkey camps'[7] (also known as 'plug and play' camps, in case we were not already aware of the linkages between Burning Man and the tech scene) and restated their commitment to the festival's principles. Much like with the word 'sharing', there is a struggle under way over the meaning and feeling of Burning Man.

As shown throughout this book, the concept of sharing does not always live up to its promise. Sometimes, it really makes no promise, or only a rather perfunctory one. For instance, when I set up printer sharing on my home network I experience none of the values associated with sharing; I am not meant to experience them; and I certainly do not feel moved to declare that 'it isn't really sharing'. The promise of sharing, in other words, is context dependent. Notably, though, the broad context in which sharing became caring and associated with rainbows is also the broad context that threatens to make sharing exploitative and alienating. This is the lesson I take from Kozinets' (2002) and Turner's (2009) analyses of Burning Man. Pragmatically speaking, sharing represents an alternative to capitalism, and at the same time it is the mode of our participation in its cutting edge. Sharing is the name of the type of communication I use when I want to discuss personal matters of import with my significant other; but when I do that with the assistance of communication technologies, my communication is commodified. This is the paradox of 'sharing' (and sharing) today.

'Sharing' combines digital culture with therapy culture with an attractive model of economics. On the one hand, the word is just a useful device, pointing us in interesting directions as we try to understand how we live today. On the other hand, the linkages between these social spheres are constituted not only by the word 'sharing', but also by the values associated with it, which are picked up and moved around from one sphere to the other. Hence, 'sharing' is not just a heuristic, but a metaphor we live by.

The notion of sharing promises a set of values: it will bring us closer to our fellow humans; it will close social gaps; it will serve as a restorative from the excesses of capitalist society. But the practices that are called 'sharing' today do not necessarily bring us closer to fulfilling its promise, leading people to say that they are 'not really sharing'. If we do not believe we can share, though, or if we think that sharing is being taken away from us, then how can we carry on? If the promise is extinguished, what are we to do? Maybe just despair, which is why various writers wish to protect 'sharing', which they do through talk of real, or pure, or genuine sharing. But the sacralization of sharing is only comprehensible under the conditions that, for some people, threaten to extinguish it. 'Sharing' did not have all of the positive connotations that it does today until people felt that society was taking us away from those values (which is not to say that society was taking us away from those values, or indeed that it was not). When they say, 'it's not really sharing', the 'sharing' that people say 'it' isn't is a product of the same social processes that created the 'it' and called it 'sharing'. Highlighting this kind of lay metalinguistic concern with words aligns this book with Raymond Williams' endeavours in his classic, *Keywords*. He states that language does not 'simply [reflect] the processes of society and history'; rather, he says, 'some important social and historical processes occur *within* language' (Williams, 1976: 22). This is what I am intimating when I claim that the same 'social and historical processes' that gave 'sharing' its present meanings are the same processes behind the complaints about its use. Notions about 'pure' or 'true' sharing, that is, are *not* antecedent to the capitalist context in which they are deployed. It is not only that they are a response to perceptions of an exploitative and alienating social order (which parts of the 'sharing economy' are sometimes seen as sharpening); they also draw on conceptions about the self and its relations with others – with particular regard to the place of communication in those relations – that are distinctly modern in character.

In the Introduction to this book I said that this is the age of sharing. By this I mean that social media and computer-mediated communication more generally are, for most of us, an inescapable part of daily life and comprise a large chunk of how we interact with the world. I also mean to point to the rise of sharing economies, be they of production or consumption, and be they for-profit or not-for-profit. And I also mean to highlight the type of talk on which our closest relationships are based, and through that to imply a particular type of self.

In addition to this, throughout the book we have seen when the practices called 'sharing' came to be thus described. This is what enables me to assert that the constellation of meanings contained with 'sharing' today is new. Equally new is the resistance to calling some of these practices 'sharing'.

This book could not have been written ten years ago, and probably not five years ago either. In the 1930s, the Oxford Group developed sharing as a type of intimate talk, which became institutionalized in AA, and thence throughout the field of support groups. But at that time 'sharing' had no technological connotations and implied nothing about the redistribution of material resources. In the 1960s and 1970s, the American counterculture was big on sharing, seeing it as a form of authentic self-expression and communication. Also, this time an egalitarian approach to material goods was implied. At the same time, sharing was becoming a key concept in computing, but in a way that was devoid of any normative aspect. Even when the power of computers for subversive social change was posited, the notion of sharing was not used in a way that united the programmes of the counterculture and progress in computer design and networking. People were using computer networks to communicate with one another from the outset, but it was not until the meteoric rise of social network sites in the early to mid-2000s that 'sharing' became the dominant term to describe computer-mediated communication and, when it did, it brought with it the meanings of sharing as authentic communication. When the sharing economy as we know it today burst on the scene in around 2010, it combined the digital and communicative senses of the word with a challenge to the materialistic belief in the need to own as much as possible. It is only now that 'sharing' ignites these semantic fields – the digital, the emotional and the economic – at the same time, and in doing so both captures and constructs our contemporary, digital lives. With apologies once more to Williams, the word 'sharing' today conveys a 'structure of feeling', a 'particular quality of social experience and relationship, historically distinct from other particular qualities, which gives the sense of a generation or period' (Williams, 1977: 131).

By excavating 'sharing', I hope to have shown the 'implicit connections' that Raymond Williams elucidated for his keywords (Williams, 1976). Others have undoubtedly written more insightfully than I have here about social media, or the therapeutic discourse, or about P2P economies. However, I have brought them together through the concept of sharing, and I have demonstrated how 'sharing' gives meaning to each of those spheres both individually and, through the transport of meaning from one to the other, collectively. The outcome of this exercise is thus greater than the sum of its parts: sharing emerges as a complex and contradictory set of practices and meanings through which we can read and make sense of large swathes of contemporary society; it is also a normative yardstick by which we evaluate the way we live. The different meanings of sharing with their different histories; the way that sharing is both a prism through which to understand our lives today and a programme by which to live our lives; the debates over what sharing really is that feel timeless but that are actually surprisingly new: all these have coalesced *now* to make this the age of sharing.

Notes

Chapter 1   Introduction

1   And see Samuel 13:20: 'But all the Israelites went down to the Philistines, to sharpen every man his share, and his coulter, and his axe, and his mattock.'
2   See: <http://www.zipcar.com/about>.
3   Even today, when I set up printer sharing on my home network, I do not experience any of the positive values usually associated with sharing, and it would be slightly strange if I did.
4   Williams, of course, was not the first to try and identify key cultural concepts. Sherry Ortner, for instance, cites social anthropologist E. E. Evans-Pritchard, who said that 'the most difficult task in social anthropological field work is to determine the meanings of a few key words, upon an understanding of which the success of the whole investigation depends' (Evans-Pritchard, 1962: 80; cited in Ortner, 1973: 1338). Ortner's interest as a cultural anthropologist is in what she calls 'key symbols' (Ortner, 1973), suggesting that a concern with key elements of a culture is not restricted to words, but that it extends also to symbols that might serve as a 'cultural focus of interest' (p. 1344).
5   See: <http://www.shareable.net/blog/the-15-best-shareable-books-of-2009>.
6   Picture downloaded 21.2.2011, available on the Wayback Machine at: <http://web.archive.org/web/20110210201533/http://www.ecosharing.net>.

7   See: <http://onthecommons.org/what-we-believe>.

8   I have guest edited one, with Wolfgang Sützl, for *Information, Communication & Society* (see John and Sützl, 2016, for the introduction); Linda Price and Russell Belk have compiled one for the *Journal of the Association for Consumer Research* (see Price and Belk, 2016, for their introduction), and Felix Stalder and Wolfgang Sützl edited one for the *International Review of Ethics* (Stalder and Sützl, 2011).

9   Such as the two-day conference on the ethics of sharing in Innsbruck, Austria in 2012; a pre-conference for the International Communication Association in Seattle in 2014; a conference on sharing and sociality in England in 2015; and the 2016 Kultursymposium in Weimar. And this does not include the plethora of conferences and colloquia devoted to the sharing economy.

Chapter 2   How Sharing Became Caring

1   Sherry Ortner (1973) reminds us that Mary Douglas (1966) points out how living organisms often provide the basis for metaphors.

2   1590 Spenser, *Faerie Queene* II.x.28: In his crowne he counted her no haire, But twixt the other twaine his kingdome whole did shaire.
    1610 P. Holland, transl. W. Camden, Brit. i.641: He shared the Country among his companions.
    *a*1616 Shakespeare, *Timon of Athens* (1623) IV.ii.23: Good Fellowes all, The latest of my wealth Ile share among'st you.

3   Belk (2014) and many others make an argument of this kind in relation to the contemporary 'sharing economy'. I shall tackle this argument at length in Chapter 4.

4   *COHA* 'contains texts from fiction, popular magazines, newspapers and non-fiction books, and is balanced by genre from decade to decade' (Davies, 2012a: 121). For more on the advantages of *COHA* compared to other historical corpora, see Davies (2012b).

5   These corpora are available at: <http://corpus.byu.edu>.

6   The outlier is from a 1935 book about religion in which reference is made to 'the method of caring and sharing' as 'the method of overcoming evil'. 'The method of love', we are told, 'depends upon changing hearts and wills by some one's caring enough for the enemy or culprit to suffer with him and for

him' – the sharing, then, is of the other's suffering (Curry, 1935).

7   The reader can get an idea of common associations with the term 'sharing is caring' by conducting a Google Images search. Be prepared for puppies, children sharing ice cream and rainbows.

Chapter 3   Sharing and the Internet

1   For a transcript of the TED talk this quotation is taken from, see: <http://www.ted.com/talks/sherry_turkle_alone_together>.
2   Sharing is Daring, at: <https://www.facebook.com/notes/facebook/sharing-is-daring/2214737130>.
3   Of course, there are other narratives of the internet wherein it is and always has been psychologically and socially harmful, or a surveillance tool in the hands of the state and corporations, and so on. However, such conceptualizations of the internet are not relevant to my efforts to understand the place of 'sharing' in online contexts.
4   Again, this is not to deny the existence of a strong dystopian stream of writing about the internet. However, because I am interested here in how and when sharing became the *sine qua non* of prosocial internet use, work that has focused on the anti-sociality of the internet is out of my scope.
5   I am aware of the problematic nature of this term, including its implicit teleology and the way it ignores previous iterations of online sociability and user-generated content. I use it here to refer to that time in the history of the web that Web 2.0 generally refers to – the mid-2000s.
6   Other instances include Streeter (2011) and Flichy (2007).
7   Except in the context of virtual communities being organized around shared interests, which is quite a different usage of the word.
8   Lawrence Lessig also posed himself this question (Lessig, 2008), and we shall see how he answered it in Chapter 4.
9   This section is based on an article published in *New Media & Society* (see John, 2013a).
10 I'm all for sharing, but why the online obsession with revealing every detail of your life?, Charlie Brooker, *Guardian*, Sunday 29 January 2012. Available at: <http://www.theguardian.com/commentisfree/2012/jan/29/sharing-obsession-revealing-every-detail>.

11 Improving Sharing Through Control, Simplicity and Connection, at: <https://www.facebook.com/notes/facebook/improving-sharing-through-control-simplicity-and-connection/101470352130>.

12 Facebook Expands to Include Work Networks, at: <https://newsroom.fb.com/News/217/Facebook-Expands-to-Include-Work-Networks>.

13 The Next Step, at: <https://www.facebook.com/notes/facebook/the-next-step/2207522130>.

14 Note once again the rhyme. Sharing is Daring, at: <https://www.facebook.com/notes/facebook/sharing-is-daring/2214737130>.

15 Share is Everywhere, at: <https://www.facebook.com/notes/2215537130>.

16 Facebook® Enables Users to Share Video, Photos, News, Blogs and More From Anywhere on the Web, at: <http://blog.photobucket.com/facebook_enable/>.

17 See: <http://www.michaelgalpert.com/post/140737454/the-many-faces-of-facebook> for Facebook screenshots that document its changing tagline from 2004.

18 See Facebook's About page, at: <https://www.facebook.com/facebook/info>.

19 The sites studied were: AsianAvenue, aSmallWorld, Badoo, Bebo, BlackPlanet, Care2, Classmates, Dodgeball, Facebook, Fiverr, Flickr, Flixster, Fotolog, Friendster, Habbo, hi5, Last.fm, LinkedIn, LiveJournal, Meetup, Multiply, MyLife, Myspace, Myyearbook, Netlog, Orkut, PerfSpot, Piczo, SixDegrees, Skyrock, StumbleUpon, Tagged, Tribe.net, TwitPic, Twitter, Viadeo, WeeWorld, Windows Live Spaces, Xanga, XING, Yahoo! 360, Yfrog, YouTube and Zorpia.

20 See: <http://en.wikipedia.org/wiki/List_of_social_networking_websites>.

21 From the Wayback Machine's About page, at: <http://archive.org/about/faqs.php#The_Wayback_Machine>.

22 See: <www.flickr.com/help/general>.

23 Dates in parentheses refer to the date the site was crawled and saved by the Wayback Machine.

24 Similarly, Peters notes that 'Cooley reveals the inner affinity of the two senses of "communication": communication as transfer or transportation and as the communication of psyches' (Cooley, 1894; in Peters, 1999: 184).

25 See: <https://www.facebook.com/notes/facebook/keeping-count-of-sharing-across-the-web/165161437130>, 26 October 2009.

26   See: <http://www.bebo.com/c/about>, 31 January 2010.
27   See: <http://www.metafilter.com/95152/Userdriven-discon
     tent>.
28   Widely available online, for instance at: <http://finance.yahoo
     .com/news/mark-zuckerberg%E2%80%99s-ipo-letter--why
     -Facebook-exists.html>.
29   The HBO comedy series, *Silicon Valley*, pokes fun at this idea
     in almost every episode.
30   See: <http://www.facebook.com/about/privacy/your-info#how
     weuse>.
31   See: <http://www.google.com/intl/en/policies/privacy>.
32   See: <https://twitter.com/PRODesigns/status/62856391520356
     3521>.

Chapter 4   Sharing Economies

1   At the time of writing a federal judge in San Francisco allowed
     Uber drivers to bring a class-action lawsuit against the company.
     By the time you read this, Uber may have changed greatly from
     what it was in late 2015.
2   Perhaps, ultimately, this part of the sharing economy will turn
     us into what Gina Neff (2012) calls 'venture laborers', or will
     result in the taskification of the workforce, as suggested by
     Mary Gray (see: <http://marylgray.org/?p=357>).
3   The P2P Foundation hosts a set of very critical magazine arti-
     cles. See: <http://p2pfoundation.net/Sharing_Economy#Discus
     sion_2> for links.
4   See: <http://www.collaborativeconsumption.com>.
5   The interface offers results from 1800, but there are none
     for this particular search before the first blip seen in Figure 4.2
     (p. 73).
6   See: <http://www.bloomberg.com/bw/stories/2005-06-19/the
     -power-of-us>.
7   See: <http://money.cnn.com/magazines/business2/business2
     _archive/2004/05/01/ 368240/index.htm>.
8   See: <http://joi.ito.com/weblog/sharing-economy>.
9   In another translation, *Urgeschichte* is rendered as 'primal
     past' (Bird-David, 2005: 205).
10   The term 'hyper-individualist' is interesting. More specifically,
     the prefix, 'hyper-' is interesting. Botsman and Rogers also
     deploy it in front of 'consumerism'. The difference between
     'individualism' and 'hyper-individualism' is unclear, but it
     would seem that the 'hyper-' prefix somehow protects the

concept it prefaces, such that it is OK to grow up in an individualistic society, but not a hyper-individualistic one; consumption is OK, but not 'hyper-consumption'. 'Hyper-' here means 'too much'. I take this as indicative of the only half-hearted challenge issued to the status quo by certain sharing economy advocates.

11   See: <http://techcrunch.com/2011/11/14/why-the-collaborative-consumption-revolution-might-be-as-significant-as-the-industrial-revolution-tctv>.

12   I enclose the word 'primitive' in scare quotes to signal my awareness of its colonialist and racist implications; I assume the reader understands the kind of popular usage of the term to which I am referring.

13   This and the other newspaper articles cited below are from a corpus of newspaper articles I created that discuss collaborative consumption and that were published in major news publications between May 2010 and April 2012. At that period, the term 'collaborative consumption' had more traction than 'sharing economy'. After filtering out similar articles (such as when a newspaper runs a story that had previously been published elsewhere), a search of the LexisNexis database of 'major world publications' for 'collaborative consumption' yielded eighty-four newspaper articles. Of these, twenty-one were excluded from the analysis on the grounds that they did not discuss collaborative consumption in any substantive manner (for instance, articles of less than 150 words; notices of a public event at which collaborative consumption would be discussed; announcements of the publication of a book about collaborative consumption; errata notices; and so on). The remaining sixty-three articles were imported into the qualitative content analysis software package, Atlas.ti.

14   While Mauss does not talk about sharing, but rather gifting, he too expresses a desire to return to the 'old and elemental' (Mauss, 1966 [1925]: 66).

15   Andrew Oram (2001) dates the growth in the term's popularity to mid-2000.

16   The talk has been viewed almost 1 million times on the TED site (<http://www.ted.com/talks/rachel_botsman_the_case_for_collaborative_consumption>) and another 160,000 times on YouTube (<https://www.youtube.com/watch?v=AQa3kUJ PEko>), as of July 2016.

17   Perhaps here more than anywhere the book's Anglocentrism is felt. I simply do not know whether, when translated into other languages, the point made here is retained or lost. Having said that, it is precisely the argument of this book that the

present meanings of 'sharing' are temporally and culturally specific.

18   See also: <http://www.shareable.net/blog/Is-Social-Media -Catalyzing-Offline-Sharing-Economy>.

19   This does not commit me to an essential core meaning of 'sharing'. To accept that there might be 'sharewashing' is only to acknowledge that certain social actors have correctly identified a consensual-enough sense of sharing and that they, at this point in time, are seeking to affiliate themselves with the positive values associated today with 'sharing'.

20   A similar kind of account of disharmony and selfishness is offered in Edward Banfield's account of the 'amoral familism' of the Montegranesi in Italy (Banfield, 1958). I am grateful to Elihu Katz for putting me on to this book.

21   By way of analogy, consider Illouz's (1997) demonstration that our contemporary conception of romantic utopias emerged in relation to the market and processes of commodification and are profoundly embedded in them.

Chapter 5   Sharing Our Feelings

1   See <https://www.facebook.com/facebook> and scroll down...

2   Available at: <http://books.google.com/ngrams>.

3   I read nine texts about the Oxford Group to see how they discuss sharing (Cantril, 1941; Clark, 1951; Day, 1923; Eister, 1950; Kitchen, 1934; Leon, 1939; Russell, 1932; The Layman with a Notebook, 1933; Van Dusen, 1934).

4   Note that although Carbaugh is here discussing a type of communication, he nonetheless employs the sharing-as-distribution metaphor.

5   I would tentatively also link these views about the dangers of not sharing to the perception that there is something pathological about Facebook refuseniks, as described by Laura Portwood-Stacer (2013).

Chapter 6   Sharing Files

1   Following the closure in January 2012 of Megaupload, a massive repository for downloadable files, global internet

traffic was said to have dropped by 4 per cent. This figure was cited in a number of online news articles, but efforts to reach the source of it were unsuccessful. Since then, the rise of Netflix has taken users away from P2P file sharing, but BitTorrent use still accounts for over 5 per cent of fixed access internet use in the US, behind only Netflix, YouTube, and web surfing (Sandvine, 2014).

2   This part of the chapter is based on John (2014).

3   Of course, the copyright holders maintain that there *is* a tragedy, namely the deprivation of artists' source of income.

4   This point resonates interestingly with Grassmuck's (2012) analysis of the 'sharing turn', according to which sharing results from abundance and is not brought about by the necessities of scarcity.

5   This also recalls Eckhardt and Bardhi's (2015) critique of the sharing economy.

6   I say 'main forms of file sharing' because people also share files by email and services such as Dropbox and Google Drive, by using USB flash drives, by burning data on to discs and by lending TV box sets to friends.

7   The servers may be mirrored in different places, but this does not fundamentally undermine the point being made here.

8   This is not strictly true, as users can change their preferences to prevent any uploading from their computers whatsoever.

9   The problem for which BitTorrent was originally presented as a solution was how to reduce the burden on a central server that wanted to distribute large files. Thus, Spotify, the music-streaming service, works by accessing the music libraries of all users on the network; similarly, other media-streaming services, such as BBC's iPlayer, which enables users to watch BBC TV channels both live and as catch-up, also use peer-to-peer technologies in order to spread the burden of delivering data-heavy video content.

10   Schwarz (2014) unpacks the metaphor of leeching very elegantly: 'p2p networks can be leeched – tapped for content – much as blood banks can' (p. 159). To which we might add that, according to dated medical beliefs, leeching was beneficial to the health of the organism as a whole.

11   This threshold is usually less than 1. Sometimes it is a sliding scale. See Aitken (2012), Tables 5.1, 5.2 and 5.3 on pp. 99–100 for a summary of some different sites' ratio requirements.

12   Private trackers with SRE were not the first sociotechnical system to offer a built-in mechanism for reining in hit-and-runners. Prior to them, BBSs (bulletin board systems) on which

users could share pirated software, and FTP servers dedicated to this same purpose, were run with ratio systems. This is a largely undocumented aspect of the history of file sharing, but see <http://www.engadget.com/2013/07/31/the-sinister-side-of-the-80s-bbs>. Analysis of the discourse of sharing in those earlier systems is beyond the scope of this chapter.

13   A seedbox can be rented for around $15 per month.

14   This forum is not behind a username and password, and so I consider it public enough for study without needing to ask for permission or consent. Nonetheless, and even though users do not go by their real names, I shall not refer to their pseudoyms. On questions of the ethics of publicness in internet research, see Hine (2009).

15   Moderators from Forums 1 and 3 gave me permission to study their forums on condition of preserving the sites' anonymity.

16   The Pirate Bay is one of the most popular public trackers.

17   The open question was: 'Please give us your own comments on the topic of file-sharing, especially how the situation in your home country looks like and what you think will be the next big thing when it comes to the Internet and/or file-sharing.'

18   Because The Pirate Bay is a public tracker, the concept of share ratio is not a prominent one among its users.

19   This practice somewhat gives the lie to the idea that a downloaded file is a lost sale (for more on P2P and 'lost sales', see Sinnreich, 2013, esp. Chapter 4).

20   'Buffer' is the delta between the amount of bytes you have downloaded/uploaded.

Chapter 7   Conclusion

1   I am referring here to Facebook's widely covered decision to make profiles, statuses and photos public by default in 2009.

2   See: <www.oversharers.com>.

3   *South Park*'s scatological treatment of this kind of radical publicness is via a social media app called Shitter, 'a device that can actually take the thoughts in your head and send them directly to the internet'. Its only members are Cartman and Alec Baldwin. (See the *South Park* episode, 'Let Go, Let Gov,' season 17, episode 1, broadcast 25 September 2013.)

4   Vise and Malseed (2008) relate that the first Google doodle was a Burning Man image, posted on Google.com to let users know that the team was away at the festival.

5 Two of the '10 Principles of Burning Man' are 'Gifting' and 'Decommodification.' See: <http://burningman.org/culture/philosophical-center/10-principles>.

6 See: <http://www.msn.com/en-us/money/companies/the-billion aires-at-burning-man/ar-AA910Bw>, or search for 'luxury Burning Man camps' online.

7 See: <http://burningman.org/event/camps/turnkey-camping>.

References

Aitken, P. A. (2012). *The ambivalences of piracy: BitTorrent media piracy and anti-capitalism* (PhD), Leeds: University of Leeds.

Akera, A. (2001). Voluntarism and the fruits of collaboration: the IBM user group, Share. *Technology and Culture* 42/4: 710–36.

Albinsson, P. A., and Yasanthi Perera, B. (2012). Alternative marketplaces in the 21st century: building community through sharing events. *Journal of Consumer Behaviour* 11/4: 303–15. doi:10.1002/cb.1389

Alcott, L. M. (1873). *Work: A Story of Experience*. Boston, MA: Roberts Brothers.

Andrejevic, M. (2004). *Reality TV: The Work of Being Watched*. London: Rowman & Littlefield Publishers.

Andrews, A. (2009). Sour note as agents and artists clash over music file sharing. *The Telegraph*, 12 September. Retrieved from: <http://www.telegraph.co.uk/finance/newsbysector/mediatechnologyand telecoms/6180010/Sour-note-as-agents-and-artists-clash-over -music-file-sharing.html>.

Arceneaux, N., and Weiss, A. S. (2010). Seems stupid until you try it: press coverage of Twitter, 2006–9. *New Media & Society* 12/8: 1262–79.

Armer, P. (1956). *SHARE: A Eulogy to Cooperative Effort*. Retrieved from Santa Monica, CA: <http://www.dtic.mil/cgi-bin/GetTRDoc ?Location=U2&doc=GetTRDoc.pdf&AD=AD0605110>.

Aslama, M., and Pantti, M. (2006). Talking alone: reality TV, emotions and authenticity. *European Journal of Cultural Studies* 9/2: 167–84.

B., D. (1997). *Turning Point: A History of Early A.A.'s Spiritual Roots and Successes.* San Rafael, CA: Paradise Research Publications.

Baedeker, R. (2011). Rent my life! Get a job? Yeah, right. There's a much quicker way to make a buck when the economy's in the tank. *Newsweek*, 28 November, p. 31.

Banfield, E. C. (1958). *The Moral Basis of a Backward Society.* Chicago, IL: Free Press; Research Center in Economic Development and Cultural Change.

Barbrook, R. (1998). The hi-tech gift economy. *First Monday* 3/12. Available at: <http://firstmonday.org/ojs/index.php/fm/article/view/631>.

Barbrook, R., and Cameron, A. (1996). The Californian ideology. *Science as Culture* 6/1: 44–72.

Bardhi, F., and Eckhardt, G. M. (2012). Access-based consumption: the case of car sharing. *Journal of Consumer Research* 39/4: 881–98. doi:10.1086/666376

Bauwens, M. (2011). Understanding peer to peer as a relational dynamics. *International Review of Information Ethics* 15: 41–51.

Bays, H., and Mowbray, M. (1999). Cookies, gift-giving, and the Internet. *First Monday* 4/11. Available at: <http://firstmonday.org/ojs/index.php/fm/article/view/700/610>.

Beck, U., and Beck-Gernsheim, E. (1995). *The Normal Chaos of Love.* Cambridge, UK: Polity.

Beekhuyzen, J., and von Hellens, L. (2009). *Reciprocity and Sharing in an Underground File Sharing Community.* Paper presented at the 20th Australasian Conference on Information Systems Melbourne.

Beekhuyzen, J., von Hellens, L., and Nielsen, S. (2011). Underground online music communities: exploring rules for membership. *Online Information Review* 35/5: 699–715.

Belk, R. (2007). Why not share rather than own? *The ANNALS of the American Academy of Political and Social Science* 611/1: 126–40. doi:10.1177/0002716206298483

Belk, R. (2010). Sharing. *Journal of Consumer Research* 36/5: 715–34. doi:10.1086/612649

Belk, R. (2014). Sharing versus pseudo-sharing in Web 2.0. *The Anthropologist* 18/1: 7–23.

Bellah, R. N., Madsen, R., Sullivan, W. M., Swidler, A. and Tipton, S. M. (1985). *Habits of the Heart: Individualism and Commitment in American Life.* Berkeley, CA: University of California Press.

Benjamin, W., and Tiedemann, R. (1999). *The Arcades Project.* Cambridge, MA: Belknap Press.

Benkler, Y. (2004). Sharing nicely: on shareable goods and the emergence of sharing as a modality of economic production. *Yale Law Journal* 114/2: 273–359.

Benkler, Y. (2006). *The Wealth of Networks: How Social Production Transforms Markets and Freedom*. New Haven, CT: Yale University Press.

Benkler, Y. (2011). *The Penguin and the Leviathan: The Triumph of Cooperation over Self-Interest*, 1st edn. New York: Crown Business.

Benski, T., and Fisher, E. (eds) (2013). *Internet and Emotions*. New York: Routledge.

Billig, M., and MacMillan, K. (2005). Metaphor, idiom and ideology: the search for 'no smoking guns' across time. *Discourse & Society* 16/4: 459–80.

Bird-David, N. (2005). The property of sharing: western analytical notions, Nayaka contexts. In T. Widlok and W. G. Tadesse (eds), *Property and Equality*. New York: Berghahn Books, pp. 201–16.

Bollier, D. (2010). Is this a new version of communism? In J. Walljasper (ed.), *All that We Share: How to Save the Economy, the Environment, the Internet, Democracy, Our Communities, and Everything Else that Belongs to All of Us*. New York: New Press, pp. 11–13.

Botsman, R. (2011). School of life: ideas for modern living No. 50: collaborative consumption. *Observer*, 30 January, p. 8.

Botsman, R., and Rogers, R. (2010). *What's Mine is Yours: The Rise of Collaborative Consumption*. New York: HarperBusiness.

boyd, d., and Ellison, N. B. (2007). Social network sites: definition, history, and scholarship. *Journal of Computer Mediated Communication* 13/1: 210–30.

Boyle, J. (1996). *Shamans, Software, and Spleens: Law and the Construction of the Information Society*. Cambridge, MA: Harvard University Press.

Boyle, K. (2012). Why buy when you can share it all – for a price. *Washington Post*, 5 March, p. 1.

Brake, D. (2014). *Sharing Our Lives Online: Risks and Exposure in Social Media*. Basingstoke, UK; New York: Palgrave Macmillan.

Breyer, S. (1970). The uneasy case for copyright: a study of copyright in books, photocopies, and computer programs. *Harvard Law Review* 84/2: 281–351.

Brown, P. M. (1937). *The Venture of Belief*. New York: Fleming H. Revell Company.

Bryant, B. (2011). Don't splash the cash – Internet hiring is the new buying. *Evening Standard*, 7 February. Available at: <http://

www.standard.co.uk/lifestyle/dont-splash-the-cash-internet -hiring-is-the-new-buying-6564124.html>.

Buck-Morss, S. (1989). *The Dialectics of Seeing: Walter Benjamin and the Arcades Project*. Cambridge, MA: MIT Press.

Buckingham, D. (2006). Children and New Media. In L. A. Lievrouw and S. M. Livingstone (eds), *Handbook of New Media: Social Shaping and Consequences of ICTs (Updated Student edition)*. London; Thousand Oaks, CA: SAGE, pp. 76–91.

Buczynski, B. (2013). *Sharing is Good: How to Save Money, Time and Resources through Collaborative Consumption*. Gabriola, BC: New Society Publishers.

Burke, M. (2010). The role of sharing in social well-being. Retrieved from: <https://www.facebook.com/notes/facebook/the-role-of -sharing-in-social-well-being/382236972130>.

Butler, G. D. (1940). *Introduction to Community Recreation, Prepared for the National Recreation Assn*, 1st edn. New York: McGraw-Hill.

Cahan, A. (1917). *The Rise of David Levinsky*. New York; London: Harper & Brothers.

Cameron, D. (1995). *Verbal Hygiene*. London; New York: Routledge.

Cameron, D. (2000). *Good to Talk?: Living and Working in a Communication Culture*. London; Thousand Oaks, CA: Sage Publications.

Cammaerts, B. (2011). Disruptive sharing in a digital age: rejecting neoliberalism? *Continuum: Journal of Media & Cultural Studies* 25/01: 47–62.

Campbell-Kelly, M., and Aspray, W. (1996). *Computer: A History of the Information Machine*, 1st edn. New York: Basic Books.

Cantril, H. (1941). *The Psychology of Social Movements*. New York: John Wiley and Sons, Inc.

Caraway, B. R. (2012). Survey of file-sharing culture. *International Journal of Communication* 6: 21.

Carbaugh, D. (1988). *Talking American: Cultural Discourses on Donahue*. Norwood, NJ: Ablex Pub. Corp.

Carbaugh, D. (1989). Fifty terms for talk: a cross-cultural study. In S. Ting-Toomey and F. Korzenny (eds), *International and Intercultural Communication Annual*, vol. 13, pp. 93–120.

Carbaugh, D. (2007). Cultural discourse analysis: communication practices and intercultural encounters. *Journal of Intercultural Communication Research* 363: 167–82.

Carbaugh, D., Berry, M. and Nurmikari-Berry, M. (2006). Coding personhood through cultural terms and practices: silence and quietude as a Finnish 'natural way of being'. *Journal of Language and Social Psychology* 25/3: 203–20.

Carey, J. W. (1989). *Communication as Culture: Essays on Media and Society.* Boston, MA: Unwin Hyman Inc.

Castells, M. (2009). *Communication Power.* Oxford; New York: Oxford University Press.

Cenite, M., Wang, M. W., Peiwen, C. and Chan, G. S. (2009). More than just free content: motivations of peer-to-peer file sharers. *Journal of Communication Inquiry* 33/3: 206–21.

Ceruzzi, P. E. (1998). *A History of Modern Computing.* Cambridge, MA: MIT Press.

Chen, X., Chu, X. and Li, Z. (2012). Improving sustainability of BitTorrent darknets. *Peer-to-Peer Networking and Applications* 7/4: 539–54.

Chute, B. J. (1956). *Greenwillow.* New York: Dutton.

Cincotta, K. (2011). Borrower & lender be. *Sunday Age*, 20 February, p. 17.

Clark, W. H. (1951). *The Oxford Group, Its History and Significance.* New York: Bookman Associates.

Cole, D. (1974). Dollie Cole on breast cancer. *Saturday Evening Post* 246: 6–13.

Cole, R. E. (1983). A Japanese management import comes full circle. *The Wall Street Journal*, 22 February, p. 22.

Cooley, C. H. (1894). *The Theory of Transportation.* Baltimore, MD: American Economics Association.

Couldry, N. (2012). *Media, Society, World: Social Theory and Digital Media Practice.* Cambridge, UK; Malden, MA: Polity.

Courtney, K. (2011). Word on the street: collaborative consumption. *The Irish Times*, 26 February, p. 5.

Coyne, R. (2007). Cornucopia limited: design and dissent on the Internet. Cambridge, MA: MIT Press.

Critcher, C. (2008). Making waves: historical aspects of public debates about children and mass media. In S. M. Livingstone and K. Drotner (eds), *The International Handbook of Children, Media and Culture.* Los Angeles, CA: SAGE, pp. 91–104.

Curry, A. B. (1935). *Speaking of Religion.* New York: Charles Scribner's Sons.

Davies, M. (2012a). Expanding horizons in historical linguistics with the 400-million word. *Corpus of Historical American English. Corpora* 7/2: 121–57.

Davies, M. (2012b). Some methodological issues related to corpus-based investigations of recent syntactic changes in English. In T. Nevalainen and E. C. Traugott (eds), *The Oxford Handbook of the History of English.* Oxford: Oxford University Press, pp. 157–74.

Day, S. S. (1923). *The Principles of the Group.* Great Britain: Oxford Group.

de Saussure, F. (2011 [1916]). *Course in General Linguistics*. New York: Columbia University Press.

de Waal, F. B. M. (2009). *The Age of Empathy: Nature's Lessons for a Kinder Society*, 1st edn. New York: Harmony Books.

Deignan, A. (2005). *Metaphor and Corpus Linguistics*. Amsterdam; Philadelphia: John Benjamins Publishing.

Doolittle, J. (1865). Social life in China. *The Harpers Monthly* 31: 429–42.

Douglas, M. (1966). *Purity and Danger*. London: Routledge and Kegan Paul.

Douglas, S. J. (1987). *Inventing American Broadcasting, 1899–1922*. Baltimore, MD: Johns Hopkins University Press.

Dovey, J. (2000). *Freakshow: First Person Media and Factual Television*. London; Sterling, VA: Pluto Press.

Dror, Y. (2013). 'We are not here for the money': founders' manifestos. *New Media & Society* 17/4: 540–55.

Dubrofsky, R. E. (2007). Therapeutics of the self: surveillance in the service of the therapeutic. *Television & New Media* 8/4: 263–84.

Dubrofsky, R. E. (2011). Surveillance on reality television and Facebook: from authenticity to flowing data. *Communication Theory* 21/2: 111–29.

Dunbar Bromley, D. (1932). The Frenchwoman holds her own. *Harpers* 12: 102–12.

Eckhardt, G. M. and Bardhi, F. (2015). The sharing economy isn't about sharing at all. *Harvard Business Review*. Available at: <https://hbr.org/2015/01/the-sharing-economy-isnt-about-sharing-at-all>.

Edwards, P. N. (1996). *The Closed World: Computers and the Politics of Discourse in Cold War America*. Cambridge, MA: MIT Press.

Eggers, D. (2013). *The Circle: A Novel*. New York: Alfred A. Knopf.

Eister, A. W. (1950). *Drawing-room Conversion: A Sociological Account of the Oxford Group Movement*. Durham, NC: Duke University Press.

Ellison, N. B. and boyd, d. (2013). Sociality through social network sites. In W. H. Dutton (ed.), *The Oxford Handbook of Internet Studies*, 1st edn. Oxford: Oxford University Press, pp. 150–72.

Ellison, N. B., Steinfield, C. and Lampe, C. (2007). The benefits of Facebook 'friends': social capital and college students' use of online social network sites. *Journal of Computer Mediated Communication* 12/4: 1143–68.

Elwood, J. M. (1993). In search for God in story and time. *America* 169: 10–13.

Evans-Pritchard, E. E. (1962). *Social Anthropology and Other Essays*. New York: Free Press.

Finck, H. T. (1899). *Primitive Love and Love-stories*. New York: Charles Scribner's Sons.

Fischer, C. S. (1992). *America Calling: A Social History of the Telephone to 1940*. Berkeley, CA: University of California Press.

Fisher, E. (2015). 'You media': audiencing as marketing in social media. *Media, Culture & Society* 37/1: 50–67.

Fitzpatrick, M. A. (1987). Marriage and verbal intimacy. In V. J. Derlega and J. H. Berg (eds), *Self-disclosure: Theory, Research, and Therapy*. New York: Plenum Press, pp. 131–54.

Flichy, P. (2007). *The Internet Imaginaire*. Cambridge, MA: MIT Press.

Ford, S. M. (2011). Reconceptualizing the public/private distinction in the age of information technology. *Information, Communication & Society* 14/4: 550–67.

Frean, A. (2010). The start-up that's putting the social network generation in the driving seat. *The Times*, 30 December, p. 46.

Frick, M.-L., and Oberprantacher, A. (2011). Shared is not yet sharing, or: what makes social networking services public? *International Review of Information Ethics* 15: 17–23.

Fuchs, C. (2011). *Foundations of Critical Media and Information Studies*. New York: Routledge.

Füredi, F. (2004). *Therapy Culture: Cultivating Vulnerability in an Uncertain Age*. London; New York: Routledge.

Gansky, L. (2010). *The Mesh: Why the Future of Business is Sharing*. New York: Penguin.

Gaskins, K. (2010). *The New Sharing Economy*. Retrieved from: <http://latdsurvey.net/pdf/Sharing.pdf>.

Gates, B., Myhrvold, N. and Rinearson, P. (1995). *The Road Ahead*. New York: Viking.

Ghosh, R. A. (2005). Cooking pot markets: an economic model for the trade in free goods and services on the Internet. *First Monday*. Retrieved from: <http://www.firstmonday.org/htbin/cgiwrap/bin/ojs/index.php/fm/article/view/1516/1431>.

Giddens, A. (1992). *The Transformation of Intimacy: Sexuality, Love, and Eroticism in Modern Societies*. Stanford, CA: Stanford University Press.

Giesler, M. (2006). Consumer gift systems. *Journal of Consumer Research* 33/2: 283–90.

Giesler, M., and Pohlmann, M. (2003). The anthropology of file sharing: consuming Napster as a gift. *Advances in Consumer Research* 30: 273–9.

Gillespie, T. (2009). Characterizing copyright in the classroom: the cultural work of antipiracy campaigns. *Communication, Culture & Critique* 2/3: 274–318.

Gillespie, T. (2010). The politics of 'platforms'. *New Media & Society* 12/3: 347–64.

Golle, P., Leyton-Brown, K. and Mironov, I. (2001). *Incentives for Sharing in Peer-to-Peer Networks*. Paper presented at the Proceedings of the 3rd ACM conference on Electronic Commerce, Tampa, Florida, USA.

Goodman, J. D. (2010). Learning to share, thanks to the Web. *The New York Times*, 26 September, p. 2.

Grassmuck, V. (2012). The sharing turn: why we are generally nice and have a good chance to cooperate our way out of the mess we have gotten ourselves into. In W. Sützl, F. Stalder, R. Maier and T. Hug (eds), *Media, Knowledge and Education: Cultures and Ethics of Sharing*. Innsbruck: Innsbruck University Press, pp. 17–34.

Griffen-Foley, B. (2004). From Tit-Bits to Big Brother: a century of audience participation in the media. *Media, Culture & Society* 26/4: 533–48.

Gross, B., and Gross, R. (1974). *Will it Grow in a Classroom?* New York: Delacorte Press.

Guadamuz, A. (2002). The 'new sharing ethic' in cyberspace. *The Journal of World Intellectual Property* 5/1: 129–39.

Habermas, J. (1989). *The Structural Transformation of the Public Sphere: An Inquiry into a Category of Bourgeois Society*. Cambridge, MA: MIT Press.

Haddon, D. (2000). Spiritual practices for living. *Total Health* 22: 49–51.

Hardin, G. (1968). The tragedy of the commons. *Science* 162/859: 1243–8.

Harrington, M. (1995). Expert: protect yourself privacy floats free in cyberspace. *Denver Post*, 11 September, p. F14.

Harris, M., and Gorenflo, N. (2012). *Share or Die: Voices of the Get Lost Generation in the Age of Crisis*. Gabriola, BC: New Society Publishers.

Harvey, J., Smith, A. and Golightly, D. (2014). Giving and sharing in the computer-mediated economy. *Journal of Consumer Behaviour*. Available at: <http://onlinelibrary.wiley.com/journal/10.1002/(ISSN)1479-1838/earlyview>.

Hasko, V. (2012). Qualitative corpus analysis. In C. A. Chapelle (ed.), *The Encyclopedia of Applied Linguistics*. Blackwell Publishing Ltd, pp. 4758–64.

Hauben, M., and Hauben, R. (1997). *Netizens: On the History and Impact of Usenet and the Internet*. Los Alamitos, CA: IEEE Computer Society Press.

Hearn, A. (2010). Structuring feeling: Web 2.0, online ranking and rating, and the digital 'reputation' economy. *Ephemera* 10/3&4: 421–38.

Hermida, A. (2014). *Tell Everyone: Why We Share and Why It Matters*. Toronto: Doubleday Canada.

Hickman, L. (2011). It's good to share: collaborative consumption – the idea that we swap or share anything from clothes to cars rather than buy more – is a brilliantly simple concept. But does it work? *Guardian*, 15 June, p. 10.

Himanen, P. (2001). *The Hacker Ethic, and the Spirit of the Information Age*, 1st edn. New York: Random House.

Hine, C. (2009). How can qualitative internet researchers define the boundaries of their projects? In A. N. Markham and N. K. Baym (eds), *Internet Inquiry: Conversations about Method*. Los Angeles, CA: Sage Publications, pp. 1–20.

Ignatieff, M. (1988). Modern dying. *The New Republic* 199: 28–33.

Illouz, E. (1997). *Consuming the Romantic Utopia: Love and the Cultural Contradictions of Capitalism*. Berkeley, CA: University of California Press.

Illouz, E. (2003). *Oprah Winfrey and the Glamour of Misery: An Essay on Popular Culture*. New York: Columbia University Press.

Illouz, E. (2007). *Cold Intimacies: The Making of Emotional Capitalism*. Cambridge, UK: Polity.

Illouz, E. (2008). *Saving the Modern Soul: Therapy, Emotions, and the Culture of Self-help*. Berkeley, CA: University of California Press.

Jakobson, R. (1960). Closing statement: linguistics and poetics. In T. Sebeok (ed.), *Style and Language*. Cambridge, MA: MIT Press, pp. 350–77.

James, A., Jenks, C. and Prout, A. (1998). *Theorizing Childhood*: Cambridge, UK: Polity.

Jessup Moore, C. (1851). *Frank and Fanny: A Rural Story*. Boston, MA: Phillips, Samson & Co.

Jia, A. L., Rahman, R., Vinkó, T., Pouwelse, J. A. and Epema, D. H. (2011). *Fast Download but Eternal Seeding: The Reward and Punishment of Sharing Ratio Enforcement*. Paper presented at the Peer-to-Peer Computing (P2P), 2011 IEEE International Conference.

Jia, A. L., Rahman, R., Vinkó, T., Pouwelse, J. A. and Epema, D. H. (2013). Systemic risk and user-level performance in private P2P communities. *Parallel and Distributed Systems, IEEE Transactions on* 24/12: 2503–12.

Johansen, B., and Johansen, R. (2007). *Get There Early: Sensing the Future to Compete in the Present*. San Francisco, CA : Berrett-Koehler Publishers.

John, N. A. (2013a). Sharing and Web 2.0: the emergence of a keyword. *New Media & Society* 15/2: 167–82. doi:10.1177/1461444812450684

John, N. A. (2013b). The social logics of sharing. *The Communication Review* 16/3: 113–31. doi:10.1080/10714421.2013.807119

John, N. A. (2014). File sharing and the history of computing: or, why file sharing is called 'file sharing'. *Critical Studies in Media Communication* 31/3: 198–211. doi:10.1080/15295036.2013.824597

John, N. A. (2016). Sharing. In B. Peters (ed.), *Digital Keywords: A Vocabulary of Information Society and Culture*. Princeton, NJ: Princeton University Press.

John, N. A., and Sützl, W. (2016). The rise of 'sharing' in communication and media studies. *Information, Communication & Society* 19/4: 437–41.

Kalamar, A. (2013). Sharewashing is the new greenwashing. Retrieved from: <http://www.opednews.com/articles/1/Sharewashing-is-the-New-Gr-by-Anthony-Kalamar-130513-834.html>.

Kao, J. (2007). *Innovation Nation: How America is Losing its Innovation Edge, Why it Matters, and What We Can Do to Get it Back*. New York, NY: Simon and Schuster.

Kaplan, B. (1967). *An Unhurried View of Copyright*. New York: Columbia University Press.

Kash, I. A., Lai, J. K., Zhang, H. and Zohar, A. (2012). *Economics of BitTorrent Communities*. Paper presented at the Proceedings of the 21st international conference on World Wide Web.

Katriel, T. (1987). 'Bexibùdim!': ritualized sharing among Israeli children. *Language in Society* 16/03: 305–20.

Katriel, T. (1988). Haxlafot: rules and strategies in children's swapping exchanges. *Research on Language and Social Interaction* 22/1: 157–78.

Katriel, T., and Philipsen, G. (1981). 'What we need is communication': 'Communication' as a cultural category in some American speech. *Communication Monographs* 48/4: 301–17.

Kaune, S., Rumin, R. C., Tyson, G., Mauthe, A., Guerrero, C. and Steinmetz, R. (2010). *Unraveling BitTorrent's File Unavailability: Measurements and Analysis*. Paper presented at the Peer-to-Peer Computing (P2P), 2010 IEEE Tenth International Conference on.

Kennedy, J. (2013). Rhetorics of sharing: data, imagination and desire. In G. Lovink and M. Rasch (eds), *Unlike Us Reader: Social Media Monopolies and Their Alternatives*. Amsterdam: Institute of Network Cultures, pp. 127–36.

Kennedy, J. (2014). Sharing in networked culture: imagination, labour and desire (PhD dissertation). Melbourne, Victoria: Swinburne University.

Kennedy, J. (2016). Conceptual boundaries of sharing. *Information, Communication & Society* 19/4: 461–74. doi:10.1080/1369118X .2015.1115888

Kitchen, V. C. (1934). *I Was a Pagan*. New York; London: Harper & Brothers.

Kleiner, D. (2011). Trapped in the World Wide Web. In G. Friesinger, J. Grenzfurthner and T. Ballhausen (eds), *Mind and Matter: Comparative Approaches Towards Complexity*. Bielefeld: transcript Verlag, pp. 175–88.

Kline, R., and Pinch, T. (1996). Users as agents of technological change: the social construction of the automobile in the rural United States. *Technology and Culture* 37/4: 763–95.

Koontz, D. R. (1983). *Phantoms*. New York: Berkley Publishing Group.

Kozinets, R. V. (2002). Can consumers escape the market? Emancipatory illuminations from Burning Man. *Journal of Consumer Research* 29/1: 20–38. doi:10.1086/339919

Krantz, J. (1980). *Princess Daisy*, 1st edn. New York: Crown Publishers.

Kraut, R., Kiesler, S., Boneva, B., Cummings, J., Helgeson, V., and Crawford, A. (2001). Internet paradox revisited. *Journal of Social Issues* 58/1: 49–74.

Kraut, R., Patterson, M., Lundmark, V., Kiesler, S., Mukophadhyay, T., and Scherlis, W. (1998). Internet paradox: a social technology that reduces social involvement and psychological well-being? *American Psychologist* 53/9: 1017–31.

Kron, J. (1983). *Home-psych: The Social Psychology of Home and Decoration*. New York: Potter.

Kropotkin, P. A. (1902). *Mutual Aid: A Factor of Evolution*. London: Heinemann.

Kutner, L. (1989). Parent & child. *New York Times*, 28 December, p. C8.

Ladd, D. (1982). Home recording and reproduction of protected works. *American Bar Association Journal* 68: 42–5.

Laing, D. (1985). Music video: industrial product, cultural form. *Screen* 26/2: 78–83.

Lakoff, G., and Johnson, M. (1980). *Metaphors We Live By*. Chicago, IL: University of Chicago Press.

Lakoff, R. T. (1990). *Talking Power: The Politics of Language in Our Lives*. New York: Basic Books.

Larkin, B. (2008). *Signal and Noise: Media, Infrastructure, and Urban Culture in Nigeria*. Durham, NC: Duke University Press.

Larrimore, L., Li, J., Larrimore, J., Markowitz, D. and Gorski, S. (2011). Peer to peer lending: the relationship between language features, trustworthiness, and persuasion success. *Journal of Applied Communication Research* 39/1: 19–37. doi:10.1080 /00909882.2010.536844

Larsson, S. (2013). Copy me happy: the metaphoric expansion of copyright in a digital society. *International Journal for the Semiotics of Law-Revue internationale de Sémiotique juridique* 26/3: 615–34.

Larsson, S., Svensson, M., de Kaminski, M., Rönkkö, K. and Alkan Olsson, J. (2012). Law, norms, piracy and online anonymity – practices of de-identification in the global file sharing community. *Journal of Research in Interactive Marketing* 6/4: 260–80.

Lasch, C. (1978). *The Culture of Narcissism: American Life in an Age of Diminishing Expectations*, 1st edn. New York: Norton.

Lasén, A., and Gómez-Cruz, E. (2009). Digital photography and picture sharing: redefining the public/private divide. *Knowledge, Technology & Policy* 22/3: 205–15.

Layman with a Notebook, The. (1933). *What is the Oxford Group?* London: Oxford University Press.

Leadbeater, C. (2008). *We-think*. London: Profile.

Lears, T. J. J. (1995). *Fables of Abundance: A Cultural History of Advertising in America*. New York: Basic Books.

Lears, T. J. J. (1983). From salvation to self-realization: advertising and the therapeutic roots of the consumer culture, 1880–1930. In T. J. J. Lears and R. Whightman Fox (eds), *The Culture of Consumption: Critical Essays in American History, 1880–1980*. New York: Pantheon Book, pp. 1–38.

Leib, F. A. (1989). *The Fire Dream*. Novato, CA: Presidio Press.

Leon, P. (1939). *The Philosophy of Courage; or, The Oxford Group Way, by Philip Leon*. New York: Oxford University Press.

Lessig, L. (2008). *Remix: Making Art and Commerce Thrive in the Hybrid Economy*. New York: Penguin Press.

Levine, M. (2009). Share my Ride. *New York Times*, 16 March, p. MM36.

Levy, S. (1984). *Hackers: Heroes of the Computer Revolution*, 1st edn. Garden City, NY: Anchor Press/Doubleday.

Liebowitz, S. J. (2006). File sharing: creative destruction or just plain destruction? *Journal of Law and Economics* 49/1: 1–28.

Litman, J. (2004). Sharing and stealing. *Hastings Communications and Entertainment Law Journal* 27: 1–48.

Liu, Z., Dhungel, P., Wu, D., Zhang, C. and Ross, K. W. (2010). *Understanding and Improving Ratio Incentives in Private Communities*. Paper presented at the Distributed Computing Systems (ICDCS), 2010 IEEE 30th International Conference on.

Livingstone, S. M., and Lunt, P. K. (1994). *Talk on Television: Audience Participation and Public Debate*. London; New York: Routledge.

Loban, B. (2004). Between rhizomes and trees: P2P information systems. *First Monday* 9/10, at: <http://www.firstmonday.org/htbin/cgiwrap/bin/ojs/index.php/fm/article/view/1182/1102>.

Logie, J. (2006). *Peers, Pirates, and Persuasion: Rhetoric in the Peer-to-Peer Debates*. West Lafayette, IN: Parlor Press.

Lupton, D. (1998). *The Emotional Self*. London: Sage.

McArthur, E. (2015). Many-to-many exchange without money: why people share their resources. *Consumption Markets & Culture* 18/3: 239–56.

McCarthy, M. (1955). *A Charmed Life*, 1st edn. New York: Harcourt.

McGlone, M. S., and Tofighbakhsh, J. (2000). Birds of a feather flock conjointly (?): Rhyme as reason in aphorisms. *Psychological Science* 11/5: 424.

McGraw, E. (2006). 1899, from Ain't we got fun. *The Southern Review* 42: 520–9.

Macken, D. (2011). Sharing is the new shopping. *Australian Financial Review*, 5 March, p. 22.

Maloney, L. (1983). A new understanding about death. *US News & World Report* 95: 62–5.

Mansell, R. (2012). *Imagining the Internet: Communication, Innovation, and Governance*. Oxford: Oxford University Press.

Mapstone, B., and Bernstein, M. I. (1980). The founding of SHARE: NCC '80 Pioneer Day. *Annals of the History of Computing* 2: 363–72.

Marks, S. R. (1986). *Three Corners: Exploring Marriage and the Self*. Lexington, MA: Lexington Books.

Martin, M. (1991). *Hello, Central?: Gender, Technology, and Culture in the Formation of Telephone Systems*. Montreal: McGill-Queen's Press-MQUP.

Marwick, A. E. (2012). The public domain: social surveillance in everyday life. *Surveillance & Society* 9/4: 378–93.

Mateus, S. (2010). Public intimacy. *Sphera Pública* 10: 57–70.

Mauss, M. (1966 [1925]). *The Gift; Forms and Functions of Exchange in Archaic Societies*. London: Cohen & West Ltd.

Mead, M. (1953). Modern marriage. *The Nation* 177: 348–50.

Meikle, G. (2016). *Social Media: Communication, Sharing and Visibility*. New York: Routledge.

Melville, H. (1851). *Moby-Dick, or, the Whale*. New York, London: Harper & Brothers; Richard Bentley.

Michel, J.-B. et al. (2011). Quantitative analysis of culture using millions of digitized books. *Science* 331/6014: 176–82.

Miller, V. (2008). New media, networking and phatic culture. *Convergence: The International Journal of Research into New Media Technologies* 14/4; 387–400.

Minar, N., and Hedlund, M. (2001). A network of peers: peer-to-peer models through the history of the internet. In A. Oram (ed.), *Peer-to-Peer: Harnessing the Benefits of a Disruptive Technology*. Beijing; Sebastopol, CA: O'Reilly, pp. 3–20.

Mirghani, S. (2011). The war on piracy: analyzing the discursive battles of corporate and government-sponsored anti-piracy media campaigns. *Critical Studies in Media Communication* 28/2: 113–34. doi:10.1080/15295036.2010.514933.

More, P. E., and Harris, C. (1904). *The Jessica Letters: An Editor's Romance*. New York: G.P. Putnam's Sons.

Morell, M. F. (2011). The unethics of sharing: Wikiwashing. *International Review of Information Ethics* 15: 9–16.

Morgan, L. H. (1881). *Houses and House-life of the American Aborigines*. Washington DC: Contributions to North American Ethnology. Department of the Interior, US Geographical and Geological Survey of the Rocky Mountain Region, vol. 4.

Morris, L. B. (1983). The war between doctors and nurses. *Good Housekeeping* 93: 169–72.

Moskowitz, E. S. (2001). *In Therapy We Trust: America's Obsession with Self-fulfillment*. Baltimore, MD: Johns Hopkins University Press.

n/a. (1846). Michelet's Life of Luther. *North American Review* 63: 433–66.

n/a. (1944). Mahatma and Viceroy. *Time* 44: 45–7.

n/a. (1999). American Craft Council Awards 1999. *American Craft* 59: 54–65.

n/a. (2010). Location, location: big ideas. *Australian Financial Review*, 9 July, p. 30.

Navarro, M. (2010). This holiday, secondhand items gain some respect. *The New York Times*, 23 December, p. 22.

Neff, G. (2012). *Venture Labor: Work and the Burden of Risk in Innovative Industries*. Cambridge, MA: MIT Press.

Negroponte, N. (1995). *Being Digital*, 1st edn. New York: Knopf.

Niles, H. E., and Niles, M. C. H. (1935). *The Office Supervisor: His Relations to Persons and to Work*. New York: J. Wiley & Sons.

Nimmer, M. B. (1969). Does copyright abridge the first amendment guarantees of free speech and press? *UCLA Law Review* 17: 1180–1204.

Nye, D. E. (1990). *Electrifying America: Social Meanings of a New Technology, 1880–1940*. Cambridge, MA: MIT Press.

O'Donnell, J. (2007). Marketers keep pace with 'tweens'. *USA Today*, 11 April, p. 1B.

Oram, A. (ed.) (2001). *Peer-to-Peer: Harnessing the Benefits of a Disruptive Technology*. Beijing; Sebastopol, CA: O'Reilly.

Ortner, S. B. (1973). On key symbols. *American Anthropologist* 75/5: 1338–46.

Ozanne, L. K., and Ballantine, P. W. (2010). Sharing as a form of anti-consumption? An examination of toy library users. *Journal of Consumer Behaviour* 9/6: 485–98.

Parker, C. (1922). *Working with the Working Woman*. New York and London: Harper & Brothers.

Patterson, L. R. (1968). *Copyright in Historical Perspective*. Nashville, TN: Vanderbilt University Press.

Pesce, M. (2012). A shared future will connect society. *Sydney Morning Herald*, 2 January, p. 11.

Peters, J. D. (1994). The gaps of which communication is made. *Critical Studies in Mass Communication* 11/2: 117.

Peters, J. D. (1999). *Speaking into the Air: A History of the Idea of Communication*. Chicago, IL: University of Chicago Press.

Pinch, T. J., and Bijker, W. E. (1987). The social construction of facts and artefacts: or how the sociology of science and the sociology of technology might benefit each other. *Social Studies of Science* 14/3: 399–441.

Pittman, B. (1988). *AA, the Way it Began*, 1st edn. Seattle, WA: Glen Abbey Books.

Plumleigh, M. (1989). Digital audio tape: new fuel stokes the smoldering home taping fire. *UCLA Law Review* 37: 733–76.

Portwood-Stacer, L. (2013). Media refusal and conspicuous non-consumption: the performative and political dimensions of Facebook abstention. *New Media & Society* 15/7: 1041–57. doi:10.1177/1461444812465139

Predelli, S. (2003). Scare quotes and their relation to other semantic issues. *Linguistics and Philosophy* 26/1: 1–28.

Price, J. A. (1975). Sharing: the integration of intimate economies. *Anthropologica* 17/1: 3–27.

Price, L. L., & Belk, R. W. (2016). Consumer Ownership and Sharing: Introduction to the Issue. *Journal of the Association for Consumer Research* 1/2: 193–7.

Putnam, R. D. (2000). *Bowling Alone: The Collapse and Revival of American Community*. New York: Simon & Schuster.

Raymond, E. S. (1996). *The New Hacker's Dictionary*, 3rd edn. Cambridge, MA: MIT Press.

Reinecke, L., and Trepte, S. (2014). Authenticity and well-being on social network sites: a two-wave longitudinal study on the effects of online authenticity and the positivity bias in SNS communication. *Computers in Human Behavior* 30: 95–102.

Rheingold, H. (1993). *The Virtual Community: Homesteading on the Electronic Frontier*. Reading, MA: Addison-Wesley Pub. Co.

Rheingold, H. (2000). *The Virtual Community: Homesteading on the Electronic Frontier*, revsd edn. Cambridge, MA: MIT Press.

Rheingold, H. (2007). *Smart Mobs: The Next Social Revolution*. New York: Basic books.

Rieff, P. (1966). *The Triumph of the Therapeutic: Uses of Faith after Freud*, 1st edn. New York: Harper & Row.

Ripeanu, M., Mowbray, M., Andrade, N. and Lima, A. (2006). Gifting technologies: a BitTorrent case study. *First Monday* 11/11. Retrieved from: <http://firstmonday.org/ojs/index.php/fm/article/view/1412/1330>.

Ripley, W. Z. (1926). Stop, look, listen: the shareholder's right to adequate information. *Atlantic Monthly* 138: 380–99.

Roberts, G. (2012). Can't afford that snazzy jacket? Then lease it; renting desirable but expensive outfits is the new way to shop. *Independent*, 22 March. Available at: <http://www.independent.co.uk/life-style/fashion/news/cant-afford-that-snazzy-jacket-then-lease-it-7555048.html>.

Rose, R. L., and Wood, S. L. (2005). Paradox and the consumption of authenticity through reality television. *Journal of Consumer Research* 32/2: 284–96.

Rosenzweig, R. (1998). Wizards, bureaucrats, warriors, and hackers: writing the history of the Internet. *American Historical Review* 103/5: 1530–52.

Roth, P. (1974). *My Life as a Man*, 1st edn. New York: Holt.

Russell, A. J. (1932). *For Sinners Only*. New York and London: Harper & Brothers.

Sandvine. (2014). *Global Internet Phenomena Report: 2H 2014*. Retrieved from: <https://www.sandvine.com/downloads/general/global-internet-phenomena/2014/2h-2014-global-internet-phenomena-report.pdf>.

Sarton, M. (1950). *Shadow of a Man*. New York: Rinehart.

Scannell, P. (2009). The question of technology. In M. Bailey (ed.), *Narrating Media History*. London; New York: Routledge, pp. 199–211.

Scholz, T. (2016). *Uberworked and Underpaid*. Cambridge, UK: Polity.

Schor, J., and Fitzmaurice, C. (2015). Collaborating and connecting: the emergence of the sharing economy. *Handbook on Research* on *Sustainable Consumption*. Cheltenham, UK: Edward Elgar, pp. 410–25.

Schwarz, J. A. (2014). *Online File Sharing: Innovations in Media Consumption*. New York: Routledge.

Searle, J. R. (1969). *Speech Acts: An Essay in the Philosophy of Language*. London: Cambridge University Press.

Shirky, C. (2008). *Here Comes Everybody: The Power of Organizing without Organizations*. New York: Penguin Press.

Shirky, C. (2010). *Cognitive Surplus: Creativity and Generosity in a Connected Age*. New York: Penguin Press.

Sigourney, L. H., and Lathrop, J. (1824). *Sketch of Connecticut, Forty Years Since*. Hartford: Oliver D. Cooke & Sons.

Singer, N. (2015). Twisting words to make 'sharing' apps seem selfless. *The New York Times*, 9 August, p. BU3.

Sinnreich, A. (2013). *The Piracy Crusade: How the Music Industry's War on Sharing Destroys Markets and Erodes Civil Liberties*. Amherst, MA: University of Massachusetts Press.

Skeggs, B. (2009). The moral economy of person production: the class relations of self-performance on 'reality' television. *The Sociological Review* 57/4: 626–44.

Slee, T. (2015). *What's Yours is Mine: Against the Sharing Economy*. New York; London: OR Books.

Sontag, S. (1978). *Illness as Metaphor*. New York: Farrar, Straus and Giroux.

Stalder, F. (1999). Beyond portals and gifts: towards a bottom-up net-economy. *First Monday* 4/1.

Stalder, F., and Sützl, W. (2011). Ethics of sharing. *International Review of Information Ethics* 15: 2.

Stefanone, M. A., Lackaff, D. and Rosen, D. (2010). The relationship between traditional mass media and 'social media': reality television as a model for social network site behavior. *Journal of Broadcasting & Electronic Media* 54/3: 508–25.

Stefansson, V. (1913). *My Life with the Eskimo*. New York: Macmillan Company.

Streeter, T. (2011). *The Net Effect: Romanticism, Capitalism, and the Internet*. New York: New York University Press.

Sweester, E. (1990). *From Etymology to Pragmatics*. Cambridge; New York: Cambridge UniversityPress.

Tapscott, D., and Williams, A. D. (2006). *Wikinomics: How Mass Collaboration Changes Everything*. New York: Portfolio Trade.

Terranova, T. (2000). Free labor: Producing culture for the digital economy. *Social Text* 18/2: 33–58.

Tetzlaff, D. (2000). Yo-ho-ho and a server of warez. In A. Herman and T. Swiss (eds), *The World Wide Web and Contemporary*

Cultural Theory: Magic, Metaphor, Power. New York: Routledge, pp. 99–126.

Thomas, F. W., and Lang, A. R. (1937). *Principles of Modern Education.* Boston, MA: Houghton Mifflin Company.

Thomas, J. (1995). *Meaning in Interaction: An Introduction to
Pragmatics.* London; New York: Longman.

Tomasello, M. (2009). *Why We Cooperate.* Cambridge, MA: MIT
Press.

Turnbull, C. M. (1961). *The Forest People.* New York: Simon and
Schuster.

Turnbull, C. M. (1972). *The Mountain People.* New York: Simon
and Schuster.

Turner, F. (2006). *From Counterculture to Cyberculture: Stewart
Brand, the Whole Earth Network, and the Rise of Digital Uto-
pianism.* Chicago, IL: University of Chicago Press.

Turner, F. (2009). Burning Man at Google: a cultural infrastructure
for new media production. *New Media & Society* 11/1–2:
73–94.

Turner, G. (2010). *Ordinary People and the Media: The Demotic
Turn.* Los Angeles, CA: SAGE.

Van Dijck, J. (2013). *The Culture of Connectivity: A Critical
History of Social Media.* Oxford; New York: Oxford University
Press.

Van Dijck, J., and Nieborg, D. (2009). Wikinomics and its discon-
tents: a critical analysis of Web 2.0 business manifestos. *New
Media & Society* 11/5: 855–74.

Van Dijk, T. A. (1993). Principles of critical discourse analysis.
Discourse & Society 4/2: 249–83.

Van Dusen, H. P. (1934). Apostle to the twentieth century: Frank
N. D. Buchman, founder of the Oxford Group movement. *Atlantic Monthly* 154/1(July): 1–16.

Veale, K. (2003). Internet gift economies: voluntary payment
schemes as tangible reciprocity. *First Monday* 8/12. Available at:
<http://firstmonday.org/ojs/index.php/fm/article/view/1101>.

Verschueren, J. (1999). *Understanding Pragmatics.* London: Oxford
University Press.

Vise, D. A., and Malseed, M. (2008). *The Google Story,* updated
edn. New York: Delacorte Press.

Voloshinov, V. N. (1986). *Marxism and the Philosophy of Language.* Cambridge, MA: Harvard University Press.

W., B. (1979). *Alcoholics Anonymous Comes of Age: A Brief
History of A.A,* 8th edn. New York: Alcoholics Anonymous
Pub.

Wallerstein, J. S., and Blakeslee, S. (1995). *The Good Marriage:
How and Why Love Lasts.* Boston, MA: Houghton Mifflin.

Walmsley, A. (2011). The age of collaboration. *Marketing*, 26 October, p. 12.

Washburn, E. (1860). *A Treatise on the American Law of Real Property*, vol. 1. Boston, MA: Little, Brown and Company.

Waters, H. F. (1983). Welcome to therapy theater. *Newsweek*, 7 November, p. 116.

Weiner, A. B. (1992). *Inalienable Possessions: The Paradox of Keeping-while-Giving*. Berkeley, CA: University of California Press.

Wellman, B., and Gulia, M. (1999). Net surfers don't ride alone: virtual communities as communities. In B. Wellman (ed.), *Networks in the Global Village*. Boulder, CO: Westview Press, pp. 331–66.

Wharton, E. (1903). *Sanctuary*. New York: Charles Scribner's Sons.

Wierzbicka, A. (1997). *Understanding Cultures through Their Key Words: English, Russian, Polish, German, and Japanese*. New York: Oxford University Press.

Wiles, K. (1952). *Teaching for Better Schools*. New York: Prentice-Hall.

Williams, R. (1976). *Keywords: A Vocabulary of Culture and Society*. Oxford: Oxford University Press.

Williams, R. (1977). *Marxism and Literature*. Oxford; New York: Oxford University Press.

Winner, L. (1980). Do artifacts have politics? *Daedalus* 109/1: 121–36.

Wittel, A. (2011). Qualities of sharing and their transformations in the digital age. *International Review of Information Ethics* 15: 3–8.

Wodak, R., and Meyer, M. (eds). (2001). *Methods of Critical Discourse Analysis*. London; Thousand Oaks, CA; New Delhi: Sage.

Woodburn, J. (1998). 'Sharing is not a form of exchange': an analysis of property-sharing in immediate-return hunter-gatherer societies. In C. M. Hann (ed.), *Property Relations: Renewing the Anthropological Tradition*. Cambridge, UK: Cambridge University Press, pp. 48–63.

Woods, T. (2001). *How I Play Golf*. New York: Warner Books.

Woodstock, L. (2014). Tattoo therapy: storying the self on reality TV in neoliberal times. *The Journal of Popular Culture* 47/4: 780–99.

Wortham, J. (2010). New type of site gives people a thrifty way to build a sense of community: online 'neighbors' can borrow, share and rent products and services. *International Herald Tribune*, 27 August, p. 15.

Wuthnow, R. (1994). *Sharing the Journey: Support Groups and America's New Quest for Community*. New York: Free Press.

Yar, M. (2008). The rhetorics and myths of anti-piracy campaigns: criminalization, moral pedagogy and capitalist property relations in the classroom. *New Media & Society* 10/4: 605–23.

Yorburg, B. (1973). *The Changing Family*. New York: Columbia University Press.

Zarsky, T. Z. (2002). Mine your own business: making the case for the implications of the data mining of personal information in the forum of public opinion. *Yale Journal of Law and Technology* 5: 1–56.

Zimmer, M. (2008). The externalities of search 2.0: the emerging privacy threats when the drive for the perfect search engine meets web 2.0. *First Monday*, 13/3), Available at: <http://www.firstmonday.org/htbin/cgiwrap/bin/ojs/index.php/fm/article/view/2136/1944>.

Index